P9-CPV-561

STUDENT SOLUTIONS MANUAL
to accompany

CALCULUS

Deborah Hughes-Hallett
Harvard University

Andrew M. Gleason
Harvard University

et al.

Prepared By: Kenny Ching
Eric Connally
Stephen A. Mallozzi
Michael Mitzenmacher
Alice H. Wang

John Wiley & Sons, Inc.
New York • Chichester • Brisbane • Toronto • Singapore

CONTENTS

Copyright © 1994 by John Wiley & Sons, Inc.

All rights reserved.

Reproduction or translation of any part of this work
beyond that permitted by Sections 107 and 108 of
the 1976 United States Copyright Act without the
permission of the copyright owner is unlawful.
Requests for permission or further information
should be addressed to the Permissions Department,
John Wiley & Sons, Inc.

ISBN 0-471-58530-0

Printed in the United States of America

10 9 8 7

CHAPTER ONE

1.1 SOLUTIONS

1. **(I)** The first graph does not match any of the given stories. In this picture, the person keeps going away from home, but his speed decreases as time passes. So a story for this might be: *I started walking to school at a good pace, but since I stayed up all night studying calculus, I got more and more tired the farther I walked.*
   (II) This graph matches (b), the flat tire story. Note the long period of time during which the distance from home did not change (the horizontal part).
   (III) This one matches (c), in which the person started calmly but sped up.
   (IV) This one is (a), in which the person forgot her books and had to return home.

5. At first, as the number of workers increases, productivity also increases. As a result, the graph of the curve goes up initially. After a certain point the curve goes downward; in other words, as the number of workers increases beyond that point, productivity decreases. This might be due either to the inefficiency inherent in large organizations or simply to workers getting in each other's way as too many are crammed on the same line.

9. The price p_1 represents the maximum price any consumer would pay for the good. The quantity q_1 is the quantity of the good that could be given away if the item were free.

13. The values $x \geq 2$ and $x \leq -2$ do not determine real values for f, because at those points either the denominator is zero or the square root is of a negative number.
    If $f(x) = 5$ then $\frac{1}{\sqrt{4-x^2}} = 5$, or $\sqrt{4 - x^2} = \frac{1}{5}$. Solving for x, we have

$$x = \pm\sqrt{\frac{99}{25}} = \pm\frac{3}{5}\sqrt{11}.$$

1.2 SOLUTIONS

1. Rewriting the equation as $y = -\frac{5}{2}x + 4$ shows that the slope is $-\frac{5}{2}$ and the vertical intercept is 4.

5. The line $y + 4x = 7$ has slope -4. Therefore the parallel line has slope -4 and equation $y - 5 = -4(x - 1)$ or $y = -4x + 9$. The perpendicular line has slope $\frac{-1}{(-4)} = \frac{1}{4}$ and equation $y - 5 = \frac{1}{4}(x - 1)$ or $y = 0.25x + 4.75$.

9. (a) Finding slope (-50) and intercept gives $q = 1000 - 50p$.
   (b) Solving for p gives $p = 20 - 0.02q$.

13. (a) Given the two points $(0, 32)$ and $(100, 212)$, and assuming the graph is a line,

$$\text{Slope} = \frac{212 - 32}{100} = \frac{180}{100} - 1.8.$$

(b)   The F-intercept is $(0, 32)$, so

$$°\text{Fahrenheit} = 1.8(°\text{Celsius}) + 32.$$

(c)   If the temperature is $20°$ Celsius, then

$$°\text{Fahrenheit} = 1.8(20) + 32 = 68°\text{Fahrenheit}.$$

(d)   If $°$Fahrenheit $=$ $°$Celsius then

$$°\text{Celsius} = 1.8°\text{Celsius} + 32$$
$$-32 = 0.8°\text{Celsius}$$
$$°\text{Celsius} = -40° = °\text{Fahrenheit}$$

17.   (a) $k = p_1 s + p_2 l$ where $s = $ # of liters of soda and $l = $ # of liters of oil.

(b)   If $s = 0$, then $l = \frac{k}{p_2}$. Similarly, if $l = 0$, then $s = \frac{k}{p_1}$. These two points give you enough information to draw a line containing the points which satisfy the equation.

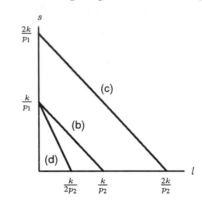

(c)   If the budget is doubled, we have the constraint: $2k = p_1 s + p_2 l$. We find the intercepts as before. If $s = 0$, then $l = \frac{2k}{p_2}$; if $l = 0$, then $s = \frac{2k}{p_1}$. The intercepts are both twice what they were before.

(d)   If the price of oil doubles, our constraint is $k = p_1 s + 2p_2 l$. Then, calculating the intercepts gives that the s intercept remains the same, but the l intercept gets cut in half. $s = 0$ means $l = \frac{k}{2p_2} = \frac{1}{2}\frac{k}{p_2}$. Therefore the maximum amount of oil you can buy is half of what it was previously.

21.   (a) $R = k(20 - H)$, where k is a positive constant. For $H > 20$, R is negative, indicating that the coffee is cooling.

(b)

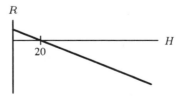

1.3 SOLUTIONS

1.

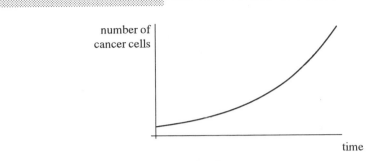

5. (a) This is the graph of a linear function, which increases at a constant rate, and thus corresponds to $k(t)$, which increases by 0.3 over each interval of 1.
    (b) This graph is concave down, so it corresponds to a function whose increases are getting smaller, as is the case with $h(t)$, whose increases are 10, 9, 8, 7, and 6.
    (c) This graph is concave up, so it corresponds to a function whose increases are getting bigger, as is the case with $g(t)$, whose increases are 1, 2, 3, 4, and 5.

9. Each increase of 1 in t seems to cause $g(t)$ to decrease by a factor of 0.8, so we expect an exponential function with base 0.8. To make our solution agree with the data at $t = 0$, we need a coefficient of 5.50, so our completed equation is
$$g(t) = 5.50(0.8)^t.$$

13. The difference, D, between the horizontal asymptote and the graph appears to decrease exponentially, so we look for an equation of the form
$$D = D_0 a^x$$
where $D_0 = 4 =$ difference when $x = 0$. Since $D = 4 - y$, we have
$$4 - y = 4a^x \quad \text{or} \quad y = 4 - 4a^x = 4(1 - a^x)$$
The point $(1, 2)$ is on the graph, so $2 = 4(1 - a^1)$, giving $a = \frac{1}{2}$.
Therefore $y = 4(1 - (\frac{1}{2})^x) = 4(1 - 2^{-x})$.

17. The doubling time is approximately 2.3. For example, the population is 20,000 at time 3.7, 40,000 at time 6, and 80,000 at time 8.3.

21. (a) Compounding 33% interest 12 times should be the same as compounding the yearly rate R once, so we get
$$\left(1 + \frac{R}{100}\right)^1 = \left(1 + \frac{33}{100}\right)^{12}$$
Solving for R, we obtain $R = 2963.51$. The yearly rate, R, is 2963.51%.

(b) The monthly rate r satisfies

$$\left(1 + \frac{4.6}{100}\right)^1 = \left(1 + \frac{r}{100}\right)^{12}$$

$$1.046^{\frac{1}{12}} = 1 + \frac{r}{100}$$

$$r = 100(1.046^{\frac{1}{12}} - 1) = 0.3755.$$

The monthly rate is 0.3755%.

1.4 SOLUTIONS

1. (a) $8^{2/3} = (8^{1/3})^2 = 2^2 = 4.$

   (b) $9^{-(3/2)} = (9^{1/2})^{(-3)} = 3^{(-3)} = \frac{1}{3^3} = \frac{1}{27}.$

5.

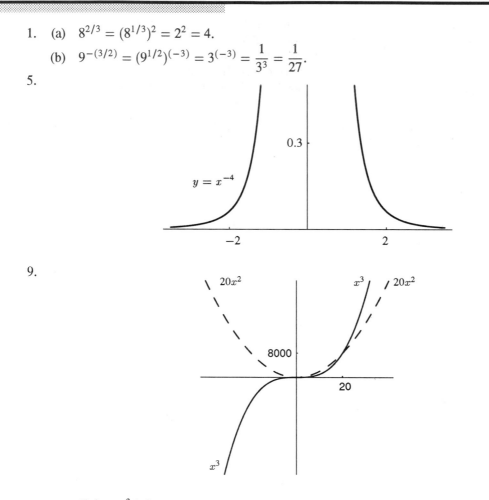

9.

$f(x) = x^3$ is larger as $x \to \infty$.

13. They are clearly equal at $x = 4$, and for all $x > 4$ the exponential function is greater. For very negative values of x, the power function is quite large, while the exponential is going to zero. The more difficult region is $-2 < x < 4$. For this region, we take a closer look at the two functions. See Figure 1.1. In this close up, it is easier to see that the exponential is greater between about -0.7 and 2. As we saw above, the exponential is also greater for $x > 4$.

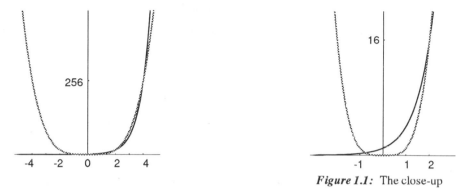

Figure 1.1: The close-up

17. $h(t)$ cannot be of the form at^2 or bt^3 since $h(0.0) = 2.04$. Therefore $h(t)$ must be the exponential, and we see that the ratio of successive values of h is approximately 1.5. Therefore $h(t) = 2.04(1.05)^t$. If $g(t) = at^2$, then $a = 3$ since $g(1.0) = 3.00$. However, $g(2.0) = 24.00 \neq 3 \cdot 2^2$. Therefore $g(t) = bt^3$, and using $g(1.0) = 3.00$, we obtain $g(t) = 3t^3$. Thus $f(t) = at^2$, and since $f(2.0) = 4.40$, we have $f(t) - 1.1t^2$.

1.5 SOLUTIONS

1. $f^{-1}(75)$ is the length of the column of mercury in the thermometer when the temperature is 75°F.
5. Not invertible, since it costs the same to mail a 50-gram letter as it does to mail a 51-gram letter.
9.

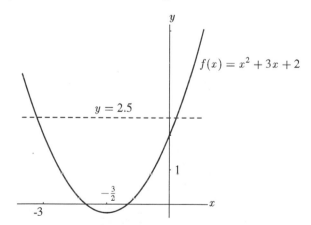

Since a horizontal line cuts the graph of $f(x) = x^2 + 3x + 2$ two times, f is not invertible.

13. (a) For each 2.2 pounds of weight the object has, it has 1 kilogram of mass, so the conversion formula is

$$k = f(p) = \frac{1}{2.2}p.$$

(b) The inverse function is

$$p = 2.2k,$$

and it gives the weight of an object in pounds as a function of its mass in kilograms.

1.6 SOLUTIONS

1.

TABLE 1.1

x	1	2	3	4	5	6	7	8	9	10
$f(x)$	0	0.30	0.48	0.60	0.70	0.78	0.85	0.90	0.95	1.00
$g(x)$	1.00	1.41	1.73	2.00	2.24	2.45	2.65	2.83	3.00	3.16

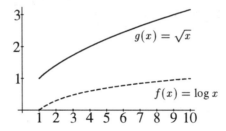

5. $t = \dfrac{\log 2}{\log 1.02} \approx 35.003.$

9. $t = \dfrac{\log \left(\frac{P}{P_0}\right)}{\log a} = \dfrac{\log P - \log P_0}{\log a}.$

13. $\log(10^{x+7}) = x + 7.$

17. $10^{-(\log B)/2} = \left[10^{\log B}\right]^{-\frac{1}{2}} = \dfrac{1}{\sqrt{B}}.$

21. If $p(t) = (1.04)^t$, then, for p^{-1} the inverse of p, we should have

$$(1.04)^{p^{-1}(t)} = t,$$
$$p^{-1}(t)\log(1.04) = \log t,$$
$$p^{-1}(t) = \frac{\log t}{\log(1.04)} \approx 58.708 \log t.$$

25. (a) Let r be the interest rate expressed as a decimal, so $r = i/100$. Then $2 = (1 + r)^D$ so $\log 2 = \log[(1 + r)^D] = D \log(1 + r)$ so

$$D = \frac{\log 2}{\log(1 + r)}$$

For $r = 0.02$, $D = 35.0$ years
For $r = 0.03$, $D = 23.4$ years
For $r = 0.04$, $D = 17.7$ years
For $r = 0.05$, $D = 14.2$ years

(b) If $D = 70/i$, then $70/2 = 35.0, 70/3 = 23.3, 70/4 = 17.5$, and $70/5 = 14.0$, showing pretty good agreement.

1.7   SOLUTIONS

1.

TABLE 1.2

x	1	2	3	4	5	6	7	8	9	10
$f(x)$	0	0.30	0.48	0.60	0.70	0.78	0.85	0.90	0.95	1.00
$g(x)$	0	0.69	1.10	1.39	1.61	1.79	1.95	2.08	2.20	2.30

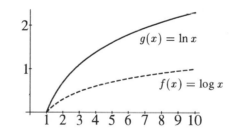

5. $e^{\ln(1/\sqrt{e})} = \frac{1}{\sqrt{e}}$

9. $\ln(e^2 \ln e) = \ln(e^2) = 2$

13. $\ln(10^{x+3}) = \ln(5e^{7-x})$
$(x + 3) \ln 10 = \ln 5 + (7 - x) \ln e$
$2.303(x + 3) = 1.609 + (7 - x)$
$\quad\quad 3.303x = 1.609 + 7 - 2.303(3)$
$\quad\quad\quad\quad x = 0.515$

17. $P = 79(e^{-2.5})^t = 79(0.0821)^t$. Exponential decay because $-2.5 < 0$ or $0.0821 < 1$.

21. We want $0.2^t = e^{kt}$ so $0.2 = e^k$ and $k = \ln 0.2 = -1.6094$. Thus $P = 5.23e^{-1.6094t}$.

25.   (a)   The quantity $\dfrac{\ln x}{\log x}$ remains constant, about $\ln 10 \approx 2.30$.

(b)   You see a horizontal line, $y \approx 2.30$, for $x > 0$. The line does not extend left of the y-axis. Thus, the function is constant for positive values of x.

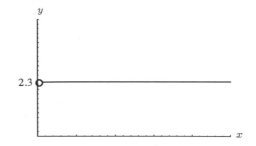

29.

$$\frac{c}{b} = e^{(-\gamma t/n)+\alpha t} = e^{t\left(\alpha - \frac{\gamma}{n}\right)}$$

$$\ln \frac{c}{b} = t\left(\alpha - \frac{\gamma}{n}\right)$$

$$t = \frac{\ln \frac{c}{b}}{\alpha - \frac{\gamma}{n}}, \quad \text{since } \alpha \neq \frac{\gamma}{n}.$$

33.   (a)   The pressure P at 6,198 meters is given in terms of the pressure P_0 at sea level to be

$$P = P_0 e^{-1.2 \times 10^{-4}h}$$
$$= P_0 e^{(-1.2 \times 10^{-4})6198}$$
$$= P_0 e^{-0.74376}$$
$$= 0.47532 P_0 \quad \text{or about 47.5\% of sea level pressure.}$$

(b)   At $h = 12,000$ meters, we have

$$P = P_0 e^{-1.2 \times 10^{-4}h}$$
$$= P_0 e^{(-1.2 \times 10^{-4})12,000}$$
$$= P_0 e^{-1.44}$$
$$= 0.2369 P_0 \quad \text{or about 24\% of sea level pressure.}$$

37.   If Q is the amount of strontium-90 which remains at time t, and Q_0 is the original amount, then

$$Q = Q_0 e^{-0.0247t}.$$

So after 100 years,

$$Q = Q_0 e^{-0.0247 \cdot 100}$$

and

$$\frac{Q}{Q_0} = e^{-2.47} \approx 0.0846$$

so about 8.46% of the strontium-90 would remain.

Note: If you assume that 2.47% is the *annual* rate, rather than the continuous rate, the answer is not very different:

$$Q = Q_0(1 - 0.0247)^{100} \quad \text{giving} \quad \frac{Q}{Q_0} \approx 0.082 \quad \text{or} \quad 8.2\%.$$

1.8 SOLUTIONS

1. For $20 \leq x \leq 100$, $0 \leq y \leq 1.2$, this function looks like a horizontal line at $y = 1.0725\ldots$ (In fact, the graph approaches this line from below.) Now, $e^{0.07} \approx 1.0725$, which strongly suggests that, as we already know, As $x \to \infty$, $\left(1 + \frac{0.07}{x}\right)^x \to e^{0.07}$.

5. $e^{0.06} = 1.0618365$, so the effective annual rate $\approx 6.18365\%$.

9. (a) Using the formula $A = A_0(1 + \frac{r}{n})^{nt}$, we have $A = 10^6(1 + \frac{1}{12})^{12} \approx 10^6(2.61303529) \approx 2{,}613{,}035$ zaïre after one year.

   (b) (i) Compounding daily, $A = 10^6(1 + \frac{1}{365})^{365} \approx 10^6(2.714567) \approx 2{,}714{,}567$ zaïre

      (ii) Compounding hourly, $A = 10^6(1 + \frac{1}{8760})^{8760} \approx 10^6(2.7181267) \approx 2{,}718{,}127$ zaïre

      (iii) Compounding each minute, $A = 10^6(1 + \frac{1}{525600})^{525600} \approx 10^6(2.718280) \approx 2{,}718{,}280$ zaïre

   (c) The amount does not seem to be increasing without bound, but rather it seems to level off at a value just over 2,718,000 zaïre. A close upper limit might be 2,718,300 (amounts may vary). In fact, the limit is $(e \times 10^6)$ zaïre.

1.9 SOLUTIONS

1. (a) The equation is $y = 2x^2 + 1$. Note that its graph is narrower than the graph of $y = x^2$ which appears in grey.

    (b) $y = 2(x^2 + 1)$ moves the graph up one unit and *then* stretches it by a factor of two.

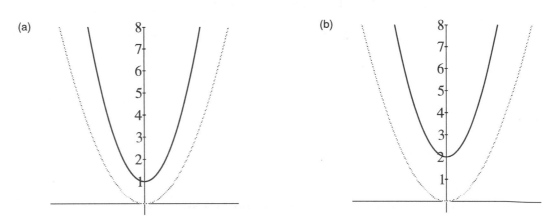

(a)   (b)

    (c) No, the graphs are not the same. Note that stretching vertically leaves any point whose y-value is zero in the same place but moves any other point. This is the source of the difference because if you stretch it first, its lowest point stays at the origin. Then you shift it up by one and its lowest point is $(0, 1)$. Alternatively, if you shift it first, its lowest point is $(0, 1)$ which, when stretched by 2, becomes $(0, 2)$.

5. (a) $f(g(t)) = f\left(\dfrac{1}{t+1}\right) = \left(\dfrac{1}{t+1} + 7\right)^2$

    (b) $g(f(t)) = g((t + 7)^2) = \dfrac{1}{(t+7)^2 + 1}$

    (c) $f(t^2) = (t^2 + 7)^2$

    (d) $g(t - 1) = \dfrac{1}{(t-1) + 1} = \dfrac{1}{t}$

9. $m(z) - m(z - h) = z^2 - (z - h)^2 = 2zh - h^2.$

13. $f(x) = x^3, \quad g(x) = \ln x.$

17.

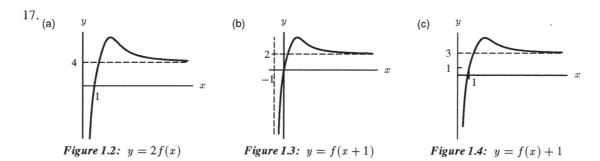

(a)

Figure 1.2: $y = 2f(x)$

(b)

Figure 1.3: $y = f(x+1)$

(c)

Figure 1.4: $y = f(x) + 1$

21.

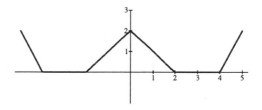

25. $f(f(1)) = f(-0.4) \approx -0.9.$

29.

TABLE 1.3

x	$f(x)$	$g(x)$	$h(x)$
-3	0	0	0
-2	2	2	-2
-1	2	2	-2
0	0	0	0
1	2	-2	-2
2	2	-2	-2
3	0	0	0

1.10 SOLUTIONS

1.

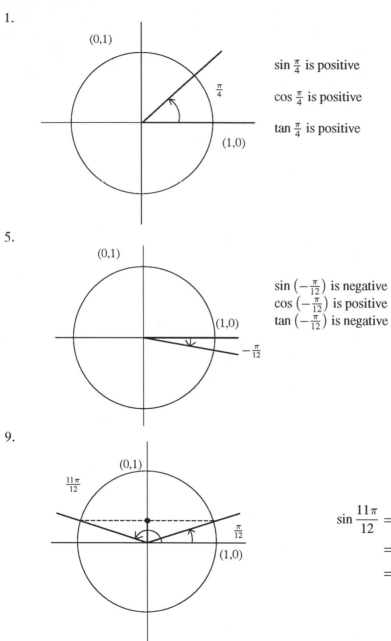

sin $\frac{\pi}{4}$ is positive

cos $\frac{\pi}{4}$ is positive

tan $\frac{\pi}{4}$ is positive

5.

sin $\left(-\frac{\pi}{12}\right)$ is negative
cos $\left(-\frac{\pi}{12}\right)$ is positive
tan $\left(-\frac{\pi}{12}\right)$ is negative

9.

$$\sin \frac{11\pi}{12} = \sin \left(\pi - \frac{\pi}{12}\right)$$
$$= \sin \frac{\pi}{12} \quad \text{(by picture)}$$
$$= 0.258.$$

13.

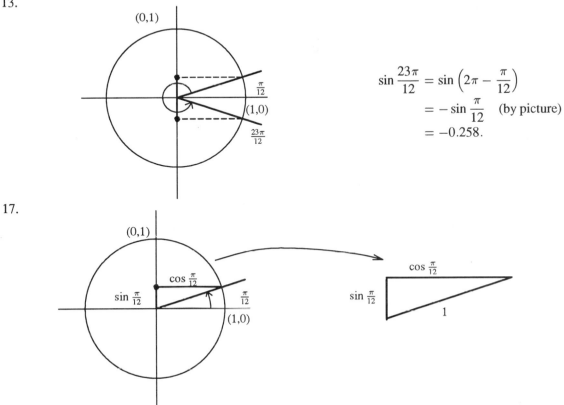

$$\sin \frac{23\pi}{12} = \sin\left(2\pi - \frac{\pi}{12}\right)$$
$$= -\sin\frac{\pi}{12} \quad \text{(by picture)}$$
$$= -0.258.$$

17.

By the Pythagorean Theorem, $(\cos \frac{\pi}{12})^2 + (\sin \frac{\pi}{12})^2 = 1^2$; so $(\cos \frac{\pi}{12})^2 = 1 - (\sin \frac{\pi}{12})^2$ and $\cos \frac{\pi}{12} = \sqrt{1 - (\sin \frac{\pi}{12})^2} = \sqrt{1 - (0.258)^2} \approx 0.966$.

We take the positive square root since by the picture we know that $\cos \frac{\pi}{12}$ is positive.

21. The moon makes one revolution around the earth in about 27.3 days, so its period is 27.3 days $\approx$ one month.

25. (a) $f(t) = -0.5 + \sin t$, $\quad g(t) = 1.5 + \sin t$, $\quad h(t) = -1.5 + \sin t$, $\quad k(t) = 0.5 + \sin t$.

   (b) $g(t) = 1 + k(t)$; $g(t) = 1.5 + \sin t = 1 + 0.5 + \sin t = 1 + k(t)$.

   (c) Since $-1 \leq \sin t \leq 1$, adding 1.5 everywhere we get $0.5 \leq 1.5 + \sin t \leq 2.5$ and since $1.5 + \sin t = g(t)$, we get $0.5 \leq g(t) \leq 2.5$. Similarly, $-2.5 \leq -1.5 + \sin t = h(t) \leq -0.5$.

29.

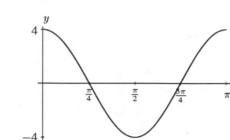

The amplitude is 4; the period is π.

33.   This graph is a sine curve with period 8π and amplitude 2, so it is given by $f(x) = 2\sin(\frac{x}{4})$.

37.   This graph is an inverted cosine curve with amplitude 8 and period 20π, so it is given by $f(x) = -8\cos(\frac{x}{10})$.

41.   (a)

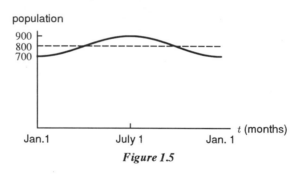

Figure 1.5

(b)   Average value of population $= \frac{700+900}{2} = 800$, amplitude $= \frac{900-700}{2} = 100$, and period $= 12$ months, so $B = 2\pi/12 = \pi/6$. Since the population is at its minimum when $t = 0$, we use a negative cosine:

$$P = 800 - 100\cos\left(\frac{\pi t}{6}\right).$$

45.   (a)

x	-1	-0.8	-0.6	-0.4	-0.2	0	0.2	0.4	0.6	0.8	1
arcsin x	-1.57	-0.93	-0.64	-0.41	-0.20	0	0.20	0.41	0.64	0.93	1.57

(b) The domain is $-1 \leq x \leq 1$.
     The range is $-\frac{\pi}{2} \leq x \leq \frac{\pi}{2}$.

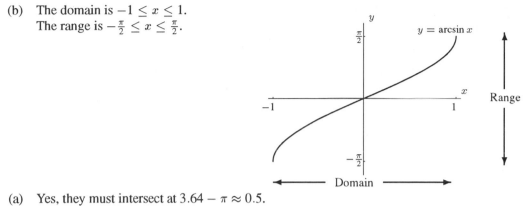

49. (a) Yes, they must intersect at $3.64 - \pi \approx 0.5$.
     (b) They also intersect at $3.64 + \pi \approx 6.78$.
     (c) $3.64 - 2\pi \approx -2.64$.

1.11 SOLUTIONS

1. (I) Degree ≥ 3, leading coefficient negative.
    (II) Degree ≥ 4, leading coefficient positive.
    (III) Degree ≥ 4, leading coefficient negative.
    (IV) Degree ≥ 5, leading coefficient negative.
    (V) Degree ≥ 5, leading coefficient positive.

5.
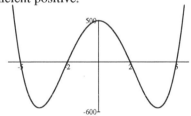

9. To find vertical asymptote(s), look at the behavior of y as x approaches a value for which the denominator is 0.

$$x^2 - 4 = 0 \quad \text{when} \quad x = \pm 2.$$

If we plug in values for x near -2 and near $+2$, we will see that

$$y \to +\infty \text{ as } x \to 2^+$$
$$y \to -\infty \text{ as } x \to 2^-$$
$$y \to -\infty \text{ as } x \to -2^+$$
$$y \to +\infty \text{ as } x \to -2^-$$

Clearly, $x = -2$ and $x = 2$ are vertical asymptotes.

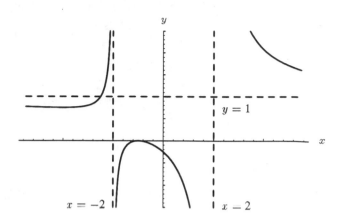

To find horizontal asymptote(s), look at the behavior of y as x goes to $+\infty$ and as x goes to $-\infty$. Note that as $x \to \pm\infty$, only the highest power of x matters, so that the $2x$, the 1, and the -4 become insignificant compared to the x^2 terms for large values of x. Thus,

$$y = \frac{x^2 + 2x + 1}{x^2 - 4} \approx \frac{x^2}{x^2} = 1$$

Clearly, $y = 1$ is a horizontal asymptote.

13. (a) If $(1, 1)$ is on the graph, we know that

$$1 = a(1)^2 + b(1) + c = a + b + c.$$

(b) If $(1, 1)$ is the vertex, then the axis of symmetry is $x = 1$, so

$$-\frac{b}{2a} = 1,$$

and thus

$$a = -\frac{b}{2}, \text{ so } b = -2a.$$

But to be the vertex, $(1, 1)$ must also be on the graph, so we know that $a + b + c = 1$. Substituting $b = -2a$, we get $-a + c = 1$, which we can rewrite as $a = c - 1$, or $c = 1 + a$.

(c) For $(0, 6)$ to be on the graph, we must have $f(0) = 6$. But $f(0) = a(0^2) + b(0) + c = c$, so $c = 6$.

(d) To satisfy all the conditions, we must first, from (c), have $c = 6$. From (b), $a = c - 1$ so $a = 5$. Also from (b), $b = -2a$, so $b = -10$. Thus the completed equation is

$$y = f(x) = 5x^2 - 10x + 6,$$

which satisfies all the given conditions.

17. (a) $f(x) = kx(x + 3)(x - 4) = k(x^3 - x^2 - 12x)$, where $k < 0$. ($k \approx -\frac{2}{9}$ if the horizontal and vertical scales are equal; otherwise one can't tell how large k is.)

   (b) This function appears to be increasing for $-1.5 < x < 2.5$, decreasing for $x < -1.5$ and for $x > 2.5$.

21. (a) Because our cubic has a root at 2 and a double root at -2, it has the form

$$y = k(x + 2)(x + 2)(x - 2).$$

Since $y = 4$ when $x = 0$,

$$4 = k(2)(2)(-2) = -8k,$$
$$k = -\frac{1}{2}.$$

Thus our equation is

$$y = -\frac{1}{2}(x + 2)^2(x - 2).$$

25. (a) $R(P) = kP(L - P)$, where k is a positive constant.

   (b)

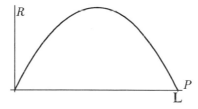

SOLUTIONS TO REVIEW PROBLEMS FOR CHAPTER ONE

1.

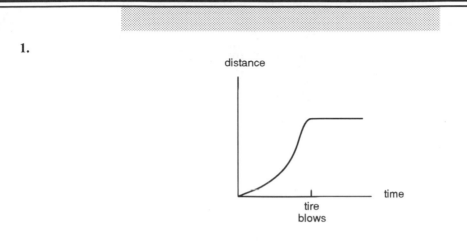

5. (a) Four zeros, at approximately $x = -4.6, 1.2, 2.7,$ and 4.1.
   (b) $f(2)$ is the y-value corresponding to $x = 2$, so $f(2)$ is about -1. Likewise, $f(4)$ is about 0.4.
   (c) Decreasing near $x = -1$, increasing near $x = 3$.
   (d) Concave up near $x = 2$, concave down near $x = -4$.
   (e) Increasing on $x < -1.5$ and on $2 < x < 3.5$.

9. (a) Advertising is generally cheaper in bulk; spending more money will give better and better marginal results initially. (Spending \$5,000 could give you a big newspaper ad reaching 200,000 people; spending \$100,000 could give you a series of TV spots reaching 50,000,000 people.)
   (b) The temperature of a hot object decreases at a rate proportional to the difference between its temperature and the temperature of the air around it. Thus, the temperature of a very hot object decreases more quickly than a cooler object. The graph is decreasing and concave up. (We are assuming that the coffee is all at the same temperature.)

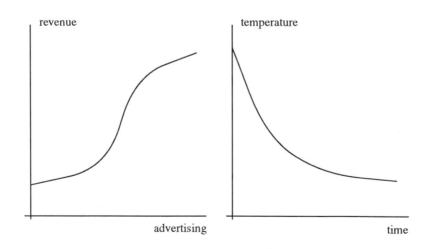

13. One possible graph is given below.

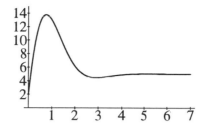

17. (a) Since $Q = 25$ at $t = 0$, we have $Q_0 = 25$ (since $e^0 = 1$). We then plug the value at $t = 1$ into the equation

$$Q = 25e^{rt}$$

to find r. Doing so, we get

$$43 = 25e^{r(1)}$$
$$\frac{43}{25} = 1.72 = e^r$$
$$\ln 1.72 = r$$
$$0.5423 \approx r.$$

And so the equation is

$$Q = 25e^{0.5423t}.$$

(b) At the time t when the population has doubled,

$$2 = e^{0.5423t}$$
$$\ln 2 = 0.5423t$$
$$t = \frac{\ln 2}{0.5423} \approx 1.3 \text{ months.}$$

(c) At the time t when the population is 1000 rabbits,

$$1000 = 25e^{0.5423t}$$
$$40 = e^{0.5423t}$$
$$\ln 40 = 0.5423t$$
$$t = \frac{\ln 40}{0.5423} \approx 6.8 \text{ months.}$$

21. To find a half-life, we want to find at what t value $Q = \frac{1}{2}Q_0$. Plugging this into the equation of the decay of plutonium-240, we have

$$\frac{1}{2} = e^{-0.00011t}$$
$$t = \frac{\ln \frac{1}{2}}{-0.00011} \approx 6,301 \text{ years.}$$

The only difference in the case of plutonium-242 is that the constant -0.00011 in the exponent is now -0.0000018. Thus, following the same procedure, the solution for t is

$$t = \frac{\ln \frac{1}{2}}{-0.0000018} \approx 385{,}000 \text{ years}.$$

25. The period T_E of the earth is (by definition!) one year or about 365.24 days (don't forget leap-years). Since the semimajor axis of the earth is 150 million km, we can use Kepler's Law to derive the constant of proportionality, k.

$$T_E = k(S_E)^{\frac{3}{2}}$$

where S_E is the earth's semimajor axis, or 150 million km.

$$365.24 = k(150)^{\frac{3}{2}}$$

$$k = \frac{365.24}{(150)^{\frac{3}{2}}} \approx 0.198.$$

Now that we know the constant of proportionality, we can use it to derive the periods of Mercury and Pluto. For Mercury,

$$T_M = (0.198)(58)^{\frac{3}{2}} \approx 87.818 \text{ days}.$$

For Pluto,

$$T_P = (0.198)(6000)^{\frac{3}{2}} \approx 92{,}400 \text{ days},$$

or (converting Pluto's period to years),

$$\frac{(0.198)(6000)^{\frac{3}{2}}}{365.24} \approx 253 \text{ years}.$$

29. Since this function has a y-intercept at $(0, 2)$, we expect it to have the form $y = 2e^{kx}$. Again, we find k by forcing the other point to lie on the graph:

$$1 = 2e^{2k}$$
$$\frac{1}{2} = e^{2k}$$
$$\ln\left(\frac{1}{2}\right) = 2k$$
$$k = \frac{\ln(\frac{1}{2})}{2} \approx -0.34657.$$

This value is negative, which makes sense since the graph shows exponential decay. The final equation, then, is

$$y = 2e^{-0.34657x}.$$

33. $x = ky(y - 4) = k(y^2 - 4y)$, where $k > 0$ is any constant.

37. (a) There are at least 3 roots in the interval: one between $x = 1$ and $x = 2$, one between $x = 2$ and $x = 3$, and one between $x = 3$ and $x = 4$.

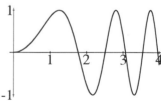

    (b) The picture of the graph to the right shows that there are a total of five roots: there are *three*, not just one, between $x = 3$ and $x = 4$.

    (c) In order, the roots are approximately $1.8, 2.5, 3.1, 3.5$, and 4.0.

    (d) $\sin x = 0$ only when x is an integer multiple of π, so $\sin(x^2) = 0$ only when t^2 is a multiple of π, say $k\pi$ where k is an integer. Thus the positive roots of $\sin(t^2)$ are $\sqrt{\pi}, \sqrt{2\pi}, \sqrt{3\pi}$, etc. The smallest positive root is therefore $\sqrt{\pi}$.

    (e) $\sqrt{\pi} \approx 1.8, \sqrt{2\pi} \approx 2.5, \sqrt{3\pi} \approx 3.1, \sqrt{4\pi} \approx 3.5, \sqrt{5\pi} \approx 4.0$.

41. (a) From the graph the period appears to be about 3:

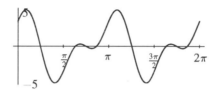

    (b) Since we are dealing with trigonometric functions, it makes sense that the actual period is π.

    (c) $\cos 2x$ is periodic with period π. $\sin 4x$ is periodic with period $\frac{\pi}{2}$, so it is also periodic with period π, since if it repeats itself after $\frac{\pi}{2}$, it will repeat itself again after π. Sums and multiples of periodic functions that have the same period are also periodic with that period, so the composition function has period π.

45. (a) $p(x) = ax^2 + c$ for any a and any c.

    (b) $p(x) = bx$ for any b.

49. (a) $S(0) = 12$ since the days are always 12 hours long at the equator.

    (b) Since $S(0) = 12$ from part (a) and the formula gives $S(0) = a$, we have $a = 12$. Since $S(x)$ must be continuous at $x = x_0$, and the formula gives $S(x_0) = a + b\arcsin(1) = 12 + b\left(\frac{\pi}{2}\right)$ and also $S(x_0) = 24$, we must have $12 + b\left(\frac{\pi}{2}\right) = 24$ so $b\left(\frac{\pi}{2}\right) = 12$ and $b = \frac{24}{\pi} \approx 7.64$.

    (c) $S(32°\,13') \approx 14.12$ and $S(46°4') \approx 15.58$.

    (d) As the graph to the right shows, $S(x)$ is not smooth. This is because above the Arctic Circle, there cannot be more than 24 hours of sunlight in a day.

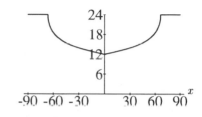

CHAPTER TWO

2.1 SOLUTIONS

1.

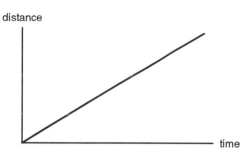

5.

$$\left(\begin{array}{c} \text{Average velocity} \\ 0 < t < 0.2 \end{array} \right) = \frac{s(0.2) - s(0)}{0.2 - 0} = \frac{0.5}{0.2} = 2.5 \text{ ft/sec.}$$

$$\left(\begin{array}{c} \text{Average velocity} \\ 0.2 < t < 0.4 \end{array} \right) = \frac{s(0.4) - s(0.2)}{0.4 - 0.2} = \frac{1.3}{0.2} = 6.5 \text{ ft/sec.}$$

A reasonable estimate of the velocity at $t = 0.2$ is the average: $\frac{1}{2}(6.5 + 2.5) = 4.5$ ft/sec.

9. $0 <$ slope at $C <$ slope at $B <$ slope of $AB < 1 <$ slope at A.

13.

TABLE 2.1

h	$(\cos h - 1)/h$
0.01	−0.005
0.001	−0.0005
0.0001	−0.00005

$$\lim_{h \to 0} \frac{\cos h - 1}{h} = 0$$

2.2 SOLUTIONS

1.   (a)

TABLE 2.2

x	1	1.5	2	2.5	3
$\log x$	0	0.18	0.30	0.40	0.48

(b)   The average rate of change of $f(x) = \log x$ between $x = 1$ and $x = 3$ is

$$\frac{f(3) - f(1)}{3 - 1} = \frac{\log 3 - \log 1}{3 - 1} \approx \frac{0.48 - 0}{2} = 0.24$$

(c)   First we find the average rates of change of $f(x) = \log x$ between $x = 1.5$ and $x = 2$, and between $x = 2$ and $x = 2.5$.

$$\frac{\log 2 - \log 1.5}{2 - 1.5} = \frac{0.30 - 0.18}{0.5} \approx 0.24$$
$$\frac{\log 2.5 - \log 2}{2.5 - 2} = \frac{0.40 - 0.30}{0.5} \approx 0.20$$

Now we approximate the instantaneous rate of change at $x = 2$ by finding the average of the above rates, i.e.

$$\left(\begin{array}{c} \text{the instantaneous rate of change} \\ \text{of } f(x) = \log x \text{ at } x = 2 \end{array}\right) \approx \frac{0.24 + 0.20}{2} = 0.22.$$

5.

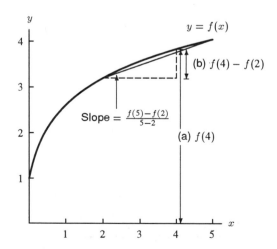

9. (a) C and D.

    (b) B and C.

    (c) A and B, and C and D.

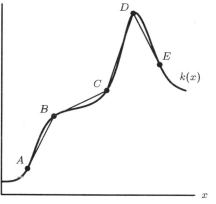

Figure 2.1

13.

$$g'(2) = \lim_{h \to 0} \frac{g(2+h) - g(2)}{h} = \lim_{h \to 0} \frac{\frac{1}{(2+h)^2} - \frac{1}{2^2}}{h}$$

$$= \lim_{h \to 0} \frac{2^2 - (2+h)^2}{2^2(2+h)^2 h} = \lim_{h \to 0} \frac{4 - 4 - 4h - h^2}{4h(2+h)^2}$$

$$= \lim_{h \to 0} \frac{-4h - h^2}{4h(2+h)^2} = \lim_{h \to 0} \frac{-4 - h}{4(2+h)^2} =$$

$$= \frac{04}{4(2)^2} = -\frac{1}{4}.$$

17.

TABLE 2.3

x	2.998	2.999	3.000	3.001	3.002
$x^3 + 4x$	38.938	38.969	39.000	39.031	39.062

We see that each x increase of 0.001 leads to an increase in $f(x)$ by about 0.031, so $f'(3) \approx \frac{0.031}{0.001} = 31$.

21. (a)

TABLE 2.4

x	$\frac{\sinh(x+0.001) - \sinh(x)}{0.001}$	$\frac{\sinh(x+0.0001) - \sinh(x)}{0.0001}$	$f'(0) \approx$	$\cosh(x)$
0	1.00000	1.00000	so 1.00000	1.00000
0.3	1.04549	1.04535	so 1.04535	1.04534
0.7	1.25555	1.25521	so 1.25521	1.25517
1	1.54367	1.54314	so 1.54314	1.54308

    (b)   It seems that they are approximately the same, i.e. the derivative of $\sinh(x) = \cosh(x)$ for $x = 0, 0.3, 0.7,$ and $1.$

25.   As h gets smaller, round-off error becomes important. When $h = 10^{-12}$, the quantity $2^h - 1$ is so close to 0 that the calculator rounds off the difference to 0, making the difference quotient 0. The same thing will happen when $h = 10^{-20}$.

2.3   SOLUTIONS

1.   The graph is that of the line $y = -2x + 2$. Its derivative is -2.

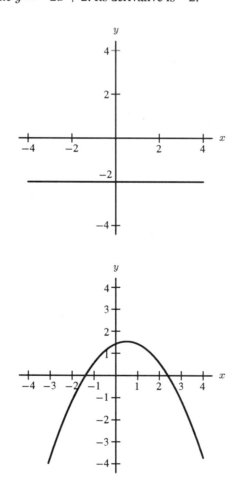

5.

9.   We know that $f'(x) \approx \dfrac{f(x+h) - f(x)}{h}$. For this problem, we'll take the average of the values obtained for $h = 1$ and $h = -1$; that's the average of $f(x+1) - f(x)$ and $f(x) - f(x-1)$ which

equals $\dfrac{f(x+1) - f(x-1)}{2}$. Thus, $f'(0) \approx f(1) - f(0) = 13 - 18 = -5$.

$f'(1) \approx [f(2) - f(0)]/2 = [10 - 18]/2 = -4$.

$f'(2) \approx [f(3) - f(1)]/2 = [9 - 13]/2 = -2$.

$f'(3) \approx [f(4) - f(2)]/2 = [9 - 10]/2 = -0.5$.

$f'(4) \approx [f(5) - f(3)]/2 = [11 - 9]/2 = 1$.

$f'(5) \approx [f(6) - f(4)]/2 = [15 - 9]/2 = 3$.

$f'(6) \approx [f(7) - f(5)]/2 = [21 - 11]/2 = 5$.

$f'(7) \approx [f(8) - f(6)]/2 = [30 - 15]/2 = 7.5$.

$f'(8) \approx f(8) - f(7) = 30 - 21 = 9$.

The rate of change of $f(x)$ is positive for $4 \le x \le 8$, negative for $0 \le x \le 3$. The rate of change is greatest at about $x = 8$.

13.

$$g'(x) = \lim_{h \to 0} \frac{g(x+h) - g(x)}{h} = \lim_{h \to 0} \frac{2(x+h)^2 - 3 - (2x^2 - 3)}{h}$$

$$= \lim_{h \to 0} \frac{2(x^2 + 2xh + h^2) - 3 - 2x^2 + 3}{h} = \lim_{h \to 0} \frac{4xh + 2h^2}{h}$$

$$= \lim_{h \to 0} (4x + 2h) = 4x$$

17.   Since $f'(x) > 0$ for $x < -1$, $f(x)$ is increasing on this interval.

Since $f'(x) < 0$ for $x > -1$, $f(x)$ is decreasing on this interval.

Since $f'(x) = 0$ at $x = -1$, the tangent to $f(x)$ is horizontal at $x = -1$.

One of many possible shapes of $y = f(x)$ is shown in Figure 2.2.

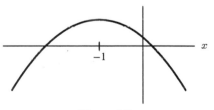

Figure 2.2

21.

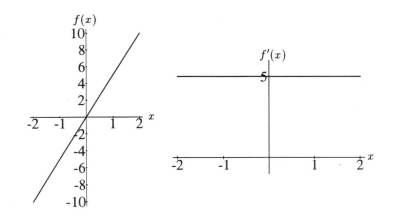

25.

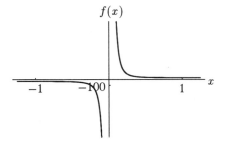

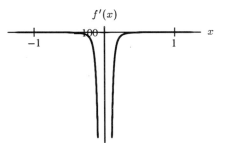

29.

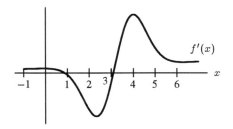

33. If $f(x)$ is even, its graph is symmetric about the y-axis. So the tangent line to f at $x = x_0$ is the same as that at $x = -x_0$ reflected about the y-axis.

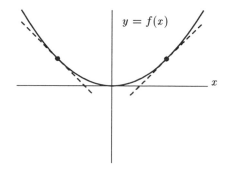

So the slopes of these two tangent lines are opposite in sign, so $f'(x_0) = -f'(-x_0)$, and f' is odd.

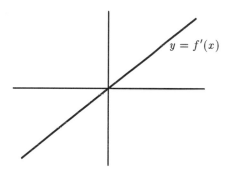

2.4 SOLUTIONS

1. (a) Velocity is zero at points A, C, F, and H.
    (b) These are points where the acceleration is zero, and hence where the particle switches from speeding up to slowing down or vice versa.

5. Units of $C'(r)$ are dollars/percent. Approximately, $C'(r)$ means the additional amount needed to pay off the loan when the interest rate is increased by 1%. The sign of $C'(r)$ is positive, because increasing the interest rate will increase the amount it costs to pay off a loan.

9. (Note that we are considering the average temperature of the yam, since its temperature is different at different points inside it.)

    (a) It is positive, because the temperature of the yam increases the longer it sits in the oven.
    (b) The units of $f'(20)$ are °F/min. $f'(20) = 2$ means that at time $t = 20$ minutes, the temperature T increases by approximately 2°F for each additional minute in the oven.

13.   (a)

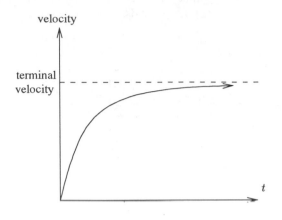

(b)   The graphs should be concave down because wind resistance decreases your acceleration as you speed up, and so the slope of the graph of velocity is decreasing.

(c)   The slope represents the acceleration due to gravity.

17.   (a)   Estimating derivatives using difference quotients (but other answers are possible):

$$P'(1900) \approx \frac{P(1910) - P(1900)}{10} = \frac{92.0 - 76.0}{10} = 1.6 \text{ million people per year}$$

$$P'(1945) \approx \frac{P(1950) - P(1940)}{10} = \frac{150.7 - 131.7}{10} = 1.9 \text{ million people per year}$$

$$P'(1990) \approx \frac{P(1990) - P(1980)}{10} = \frac{248.7 - 226.5}{10} = 2.22 \text{ million people per year}$$

(b)   The population growth was maximal somewhere between 1950 and 1960.

(c) $P'(1950) \approx \frac{P(1960) - P(1950)}{10} = \frac{179.0 - 150.7}{10} = 2.83$ million people per year, so $P(1956) \approx P(1950) + P'(1950)(1956 - 1950) = 150.7 + 2.83(6) \approx 167.7$ million people.

(d)   If the growth rate between 1990 and 2000 was the same as the growth rate from 1980 to 1990, then the total population should be about 271 million people in 2000.

2.5 SOLUTIONS

1.   (a)   increasing, concave up.

(b)   decreasing, concave down

5. (a)

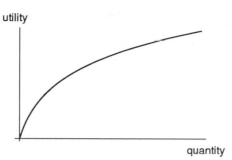

(b) As a function of quantity, utility is increasing but at a decreasing rate; the graph is increasing but concave down. So the derivative of utility is positive, but the second derivative of utility is negative.

9. (a) The EPA will say that the rate of discharge is still rising. The industry will say that the rate of discharge is increasing less quickly, and may soon level off or even start to fall.

(b) The EPA will say that the rate at which pollutants are being discharged is levelling off, but not to zero — so pollutants will continue to be dumped in the lake. The industry will say that the rate of discharge has decreased significantly.

2.6   SOLUTIONS

1. The tangent line has slope $1/4$ and goes through the point $(4, 2)$. Its equation is

$$y - 2 = \frac{1}{4}(x - 4) \qquad y = \frac{1}{4}x + 1.$$

Thus when $x = 4.007$, we have $y = \frac{1}{4}(4.007) + 1 = 2.00175$. Since 2.0017496 rounds to 2.00175, the point $(4.007, 2.0017496) \approx (4.007, 2.00175)$ lies on the tangent line.

5. (a) $f'(6.75) \approx \frac{f(7.0) - f(6.5)}{7.0 - 6.5} = \frac{8.2 - 10.3}{0.5} = -4.2.$

$f'(7.0) \approx \frac{f(7.5) - f(6.5)}{7.5 - 6.5} = \frac{6.5 - 10.3}{1.0} = -3.8.$

$f'(8.5) \approx \frac{f(9.0) - f(8.0)}{9.0 - 8.0} = \frac{3.2 - 5.2}{1.0} = -2.0.$

(b) To estimate f'' at 7, we should have values for f' at points near 7. We know from (a) that $f'(6.75) \approx -4.2$. Next, estimate $f'(7.25) \approx \frac{6.5 - 8.2}{0.5} = -3.4$. Then

$f''(7) \approx \frac{f'(7.25) - f'(6.75)}{0.5} \approx \frac{-3.4 - (-4.2)}{0.5} = 1.6.$

(c) $y - 8.2 = -3.8(x - 7)$ or $y = -3.8x + 34.8$.

(d) We may use the tangent line from (c) to approximate $f(6.8)$. In this case we get

$$y \approx -3.8x + 34.8 = 8.96.$$

[We may also estimate $f(6.8)$ by assuming that the graph of f is straight between the given points $(6.5, 10.3)$ and $(7.0, 8.2)$. This line has the equation $y = -4.2(x - 6.5) + 10.3$ and passes through $(6.8, 9.04)$, so we may estimate $f(6.8) \approx 9.04$. Here, we approximate using the secant line rather than the tangent line.]

As we can see, the two estimates are fairly close.

9. (a) The graph is a sine curve. It looks straight because the graph shows only a small part of the curve magnified greatly. The period of the sine curve is $\frac{2\pi}{0.0172} \approx 365$ (the number of days in a year).

(b) The month is March: We see that about the 21st of the month there are twelve hours of daylight and hence twelve hours of night. This phenomenon (the length of the day equaling the length of the night) occurs at the equinox, midway between winter and summer. Since the length of the days is increasing, and Madrid is in the northern hemisphere, we are looking at March, not September.

(c) The slope of the curve is found either from the graph or the formula to be about 0.04 (the rise is about 0.8 hours in 20 days or 0.04 hours/day). This means that the amount of daylight is increasing by about 0.04 hours (about $2\frac{1}{2}$ minutes) per calendar day, or that each day is $2\frac{1}{2}$ minutes longer than its predecessor.

2.7   SOLUTIONS

1. The limit appears to be 1; a graph and table of values is shown below.

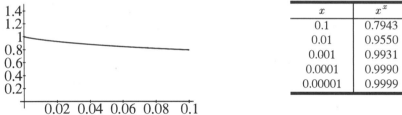

x	x^x
0.1	0.7943
0.01	0.9550
0.001	0.9931
0.0001	0.9990
0.00001	0.9999

5. For $-0.5 \le \theta \le 0.5$, $\quad 0 \le y \le 0.5$, the graph of $y = \dfrac{\theta}{\tan 3\theta}$ is shown to the right. Therefore, by tracing along the curve, we see that

$$\lim_{\theta \to 0} \frac{\theta}{\tan 3\theta} = 0.3333\ldots.$$

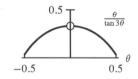

9. The only point at which g might not be continuous is at $\theta = 0$. A graph shows that

$$\lim_{\theta \to 0} \frac{\sin \theta}{\theta} = 1.$$

However

$$\lim_{\theta \to 0} \frac{\sin \theta}{\theta} = 1 \ne g(0),$$

so $g(\theta)$ is not continuous on any interval containing 0.

2.8 SOLUTIONS

1. Yes, f is differentiable at $x = 0$ (see Figure 2.3).

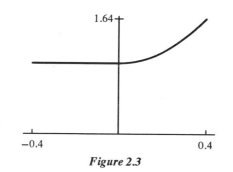

Figure 2.3

5. (a) The graph of

    $$f(x) = \begin{cases} 0 & \text{if } x < 0. \\ x^2 & \text{if } x \geq 0. \end{cases}$$

    is shown to the right. The graph is con-
    tinuous and has no vertical segments or
    corners, so $f(x)$ is differentiable every-
    where.

    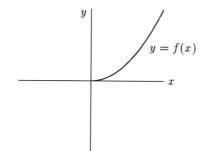

    By Example 4 on page 115,

    $$f'(x) = \begin{cases} 0 & \text{if } x < 0 \\ 2x & \text{if } x \geq 0 \end{cases}$$

    So its graph is shown to the right.

    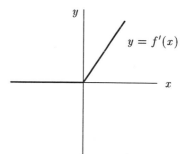

(b) The graph of the derivative has a corner at $x = 0$ so $f'(x)$ is not differentiable at $x = 0$. The graph of

$$f''(x) = \begin{cases} 0 \text{ if } x < 0 \\ 2 \text{ if } x > 0 \end{cases}$$

looks like:

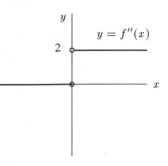

The second derivative is not defined at $x = 0$. So it is certainly neither differentiable nor continuous at $x = 0$.

SOLUTIONS TO REVIEW PROBLEMS FOR CHAPTER TWO

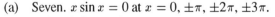

1.

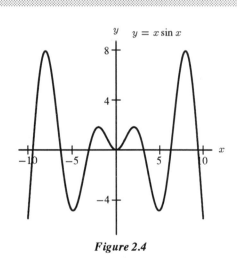

Figure 2.4

(a) Seven. $x \sin x = 0$ at $x = 0$, $\pm\pi$, $\pm 2\pi$, $\pm 3\pi$.
(b) $x \sin x$ is increasing at $x = 1$, decreasing at $x = 4$.
(c) $\dfrac{f(2) - f(0)}{(2 - 0)} = \dfrac{2 \sin 2 - 0}{2} = \sin 2 \approx 0.91$

$\dfrac{f(8) - f(6)}{(8 - 6)} = \dfrac{8 \sin 8 - 6 \sin 6}{2} \approx 4.80$. So the average rate of change over $6 \leq x \leq 8$ is greater.
(d) It's greater at $x = -9$.

5.

TABLE 2.5

x	$\ln x$	x	$\ln x$	x	$\ln x$	x	$\ln x$
0.998	−0.0020	1.998	0.6921	4.998	1.6090	9.998	2.3024
0.999	−0.0010	1.999	0.6926	4.999	1.6092	9.999	2.3025
1.000	0.0000	2.000	0.6931	5.000	1.6094	10.000	2.3026
1.001	0.0010	2.001	0.6936	5.001	1.6096	10.001	2.3027
1.002	0.0020	2.002	0.6941	5.002	1.6098	10.002	2.3028

At $x = 1$, the values of $\ln x$ are increasing by 0.001 for each increase in x of 0.001, so the derivative appears to be 1. At $x - 2$, the increase is 0.0005 for each increase of 0.001, so the derivative appears to be 0.5. At $x = 5$, $\ln x$ increases by 0.0002 for each increase of 0.001 in x, so the derivative appears to be 0.2. And at $x = 10$, the increase is 0.0001 over intervals of 0.001, so the derivative appears to be 0.1. These values suggest an inverse relationship between x and $f'(x)$, namely $f'(x) = \frac{1}{x}$.

9.

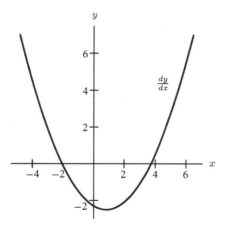

13.

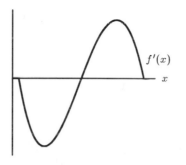

17.   (a) IV, (b) III, (c) II, (d) I, (e) IV, (f) II

21.   (a)   The population varies periodically with a period of 12 months (i.e. one year).

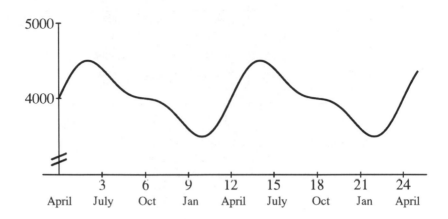

(b)   The herd is largest about June 1st when there are about 4500 deer.

(c)   The herd is smallest about February 1st when there are about 3500 deer.

(d)   The herd grows the fastest about April 1st. The herd shrinks the fastest about July 20 and again about November 15.

(e)   It grows the fastest about April 1st when the rate of growth is about 400 deer/month, i.e about 13 new fawns per day.

CHAPTER THREE

3.1 SOLUTIONS

1. (a) Lower estimate $= 60+40+25+10+0 = 135$ feet. Upper estimate $= 88+60+40+25+10 = 223$ feet.

   (b)

   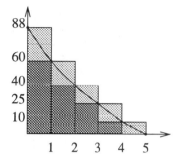

5. (a) We want the error to be less than 0.1, so take Δx such that $|f(1) - f(0)|\Delta x < 0.1$, giving

$$\Delta x < \frac{0.1}{|e^{-\frac{1}{2}} - 1|} \approx 0.25$$

   so take $\Delta x = 0.25$ or $n = 4$. Then the left sum = 0.9016, and the right sum = 0.8033, so a reasonable estimate for the area is $\frac{0.9016+0.8033}{2} = 0.8525$. Certainly 0.85 is within 0.1 of the actual answer.

   (b) Take Δx smaller. To have an error of at most E, you need Δx such that

$$|f(1) - f(0)|\Delta x < E$$

   This means

$$\Delta x < \frac{E}{|e^{-\frac{1}{2}} - 1|} \approx \frac{E}{0.39}.$$

   Using n equal subdivisions, we have

$$\Delta x = \frac{b - a}{n} = \frac{1 - 0}{n} = \frac{1}{n}.$$

   Thus, to approximate the shaded area with an error $< E$ requires $n > \frac{0.39}{E}$ subdivisions.

9. (a) An upper estimate is $9.81 + 8.03 + 6.53 + 5.38 + 4.41 = 34.16$ m/sec. A lower estimate is $8.03 + 6.53 + 5.38 + 4.41 + 3.61 = 27.96$ m/sec.

    (b) The average is $\frac{1}{2}(34.16 + 27.96) = 31.06$ m/sec. Because the graph of acceleration is concave up, this estimate is too high, as can be seen in the figure to the right. The area of the shaded region is the average of the areas of the rectangles $ABFE$ and $CDFE$.

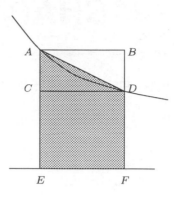

3.2 SOLUTIONS

1.
TABLE 3.1

n	2	10	50	250
Left-hand Sum	0.0625	0.2025	0.2401	0.248004
Right-hand Sum	0.5625	0.3025	0.2601	0.252004

The sums seem to be converging to $\frac{1}{4}$. Since x^3 is monotone on $[0, 1]$, the true value is between 0.248004 and 0.252004 .

5.
TABLE 3.2

n	2	10	50	250
Left-hand Sum	−0.394991	−0.0920539	−0.0429983	−0.0335556
Right-hand Sum	0.189470	0.0248382	−0.0196199	−0.0288799

There is no obvious guess as to what the limiting sum is. Moreover, since $\sin(t^2)$ is *not* monotonic on $[2, 3]$, we cannot be sure that the true value is between −0.0335556 and −0.0288799.

9. For $n = 10$, LHS ≈ 0.465 (rounding down) and RHS ≈ 0.474 (rounding up). Since the left and right sums differ by 0.009, their average must be within 0.0045 of the true value, so $\int_1^{1.5} \sin x \, dx = 0.470$ to one decimal place.

13. For $n = 30$, LHS ≈ 2.852 (rounding down) and RHS ≈ 2.919 (rounding up). Since the left and right sums differ by 0.067, their average must be within 0.0335 of the true value, so $\int_1^2 2^x \, dx = 2.886$ to one decimal place.

17. Since e^{-t^2} is an even function, $\int_{-3}^3 e^{-t^2} \, dt = 2 \int_0^3 e^{-t^2} \, dt$. This is because the integrand is symmetrical about the y-axis, and by symmetry, $\int_{-3}^0 e^{-t^2} \, dt = \int_0^3 e^{-t^2} \, dt$. Since e^{-t^2} is decreasing monotonically

on the interval $0 < t < 3$, we can use our error estimation techniques to approximate $\int_0^3 e^{-t^2}\,dt$. Since we want our error to be less than 0.05 for the interval $-3 < t < 3$, we want our error to be less than $\frac{1}{2}(0.05) = 0.025$ for the interval $0 < t < 3$. For $n = 70$, LHS ≈ 0.908 (rounding up) and RHS ≈ 0.864 (rounding down). Since the left and right sums differ by 0.044, their average must be within 0.022 of the true value, so $\int_{-3}^3 e^{-t^2}\,dt = 2\int_0^3 e^{-t^2}\,dt = 2(0.886) = 1.772$, to one decimal place.

3.3 SOLUTIONS

1. The units of measurement are foot-pounds (which are units of work).

5. Since x intercepts are $x = 0, \pi, 2\pi, \ldots$,

$$\text{Area} = \int_0^\pi \sin x\,dx = 2.00.$$

9. (a)

(b) $A_1 = \displaystyle\int_{-2}^0 f(x)\,dx = 2.667.$

$A_2 = -\displaystyle\int_0^1 f(x)\,dx = 0.417.$

So Total area $= A_1 + A_2 = 3.08.$

(c) $\displaystyle\int_{-2}^1 f(x)\,dx = A_1 - A_2 = 2.50.$

13. Average value $= \dfrac{1}{2-0}\displaystyle\int_0^2 (1+t)\,dt = \dfrac{1}{2}(4) = 2.$

17. (a) Since $f(x) = \sin x$ over $[0, \pi]$ is between 0 and 1, the average of $f(x)$ must itself be between 0 and 1. Furthermore, since the graph of $f(x)$ is concave down on this interval, the average value must be greater than the average height of the triangle shown in the figure, namely, 0.5.

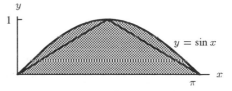

(b)   Average $= \dfrac{1}{\pi - 0} \displaystyle\int_0^\pi \sin x \, dx = \dfrac{2}{\pi} = 0.64$.

21.   (a)   For $-2 \le x \le 2$, f is symmetrical about the y-axis, so $\int_{-2}^0 f(x)\,dx = \int_0^2 f(x)\,dx$ and $\int_{-2}^2 f(x)\,dx = 2\int_0^2 f(x)\,dx$.

(b)   For any function f, $\int_0^2 f(x)\,dx = \int_0^5 f(x)\,dx - \int_2^5 f(x)\,dx$.

(c)   Note that $\int_{-2}^0 f(x)\,dx = \frac{1}{2}\int_{-2}^2 f(x)\,dx$, so $\int_0^5 f(x)\,dx = \int_{-2}^5 f(x)\,dx - \int_{-2}^0 f(x)\,dx = \int_{-2}^5 f(x)\,dx - \frac{1}{2}\int_{-2}^2 f(x)\,dx$.

3.4 SOLUTIONS

1.   Since $F(0) = 0$, $F(b) = \int_0^b f(t)\,dt$. For each b we determine $F(b)$ graphically as follows:
$F(0) = 0$
$F(1) = F(0) + \text{Area of } 1 \times 1 \text{ rectangle} = 0 + 1 = 1$
$F(2) = F(1) + \text{Area of triangle } (\frac{1}{2} \cdot 1 \cdot 1) = 1 + 0.5 = 1.5$
$F(3) = F(2) + \text{Negative of area of triangle} = 1.5 - 0.5 = 1$
$F(4) = F(3) + \text{Negative of area of rectangle} = 1 - 1 = 0$
$F(5) = F(4) + \text{Negative of area of rectangle} = 0 - 1 = -1$
$F(6) = F(5) + \text{Negative of area of triangle} = -1 - 0.5 = -1.5$
The graph of $F(t)$, for $0 \le t \le 6$, is shown in Figure 3.1.

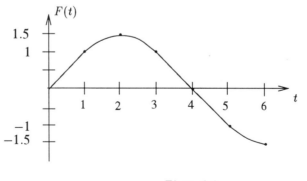

Figure 3.1

4.   The graph of $f(\theta) = F'(\theta) = \sin(\theta^2)$ is in Figure 3.2.

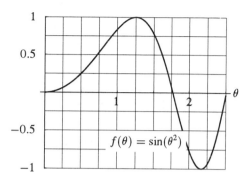

Figure 3.2

F is increasing where $F'(\theta) = f(\theta) > 0$, i.e. for $0 < \theta < \sqrt{\pi} \approx 1.77$. F is decreasing where $F'(\theta) = f(\theta) < 0$, i.e. for $1.77 < \theta \le 2.5$.

5. Using the solution to Problem 4, we see that F is increasing for $\theta < 1.77$ and F is decreasing for $\theta > 1.77$. Thus F has its maximum value when $\theta = 1.77$, and then by the Fundamental Theorem of Calculus,

$$F(1.77) - F(0) = \int_0^{1.77} f(\theta)\,d\theta = \int_0^{1.77} \sin(\theta^2)\,d\theta.$$

Since $F(0) = 0$, using Riemann sums we get

$$F(1.77) = \int_0^{1.77} \sin(\theta^2)\,d\theta = 0.89.$$

9. First rewrite each of the quantities in terms of f', since we have the graph of f'. If A_1 and A_2 are the positive areas shown in Figure 3.3:

$$f(3) - f(2) = \int_2^3 f'(t)\,dt = -A_1$$

$$f(4) - f(3) = \int_3^4 f'(t)\,dt = -A_2$$

$$\frac{f(4) - f(2)}{2} = \frac{1}{2}\int_2^4 f'(t)\,dt = -\frac{A_1 + A_2}{2}$$

Since Area $A_1 >$ Area A_2,

$$A_2 < \frac{A_1 + A_2}{2} < A_1$$

so

$$-A_1 < -\frac{A_1 + A_2}{2} < -A_2$$

and therefore

$$f(3) - f(2) < \frac{f(4) - f(2)}{2} < f(4) - f(3).$$

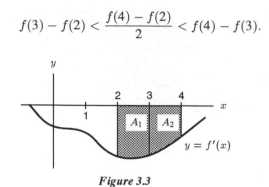

Figure 3.3

13. (a) Quantity used $= \int_0^5 f(t)\,dt$.
    (b) Using a left sum, our approximation is

$$32e^{0.05(0)} + 32e^{0.05(1)} + 32e^{0.05(2)} + 32e^{0.05(3)} + 32e^{0.05(4)} = 177.27.$$

Since f is an increasing function, this represents an underestimate.

(c) Each term is a lower estimate of one year's consumption of oil.

17.

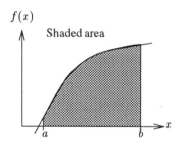

21.

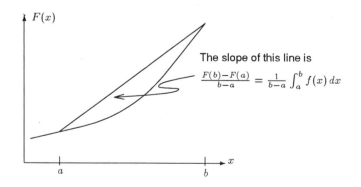

3.5 SOLUTIONS

1. For $0 \leq t \leq 10, 0 \leq y \leq 1$ the graph of $y = te^{-t}$ looks like the figure to the right.
    Therefore,

    $$\lim_{t \to \infty} te^{-t} = 0.$$

    Also, we are taking the limit of $\dfrac{t}{e^t}$ as $t \to \infty$, and as discussed in Chapter 1, Section 1.4, exponential functions grow much faster than t as $t \to \infty$.

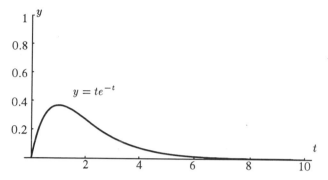

5. (a) It's true, for example, that since

    $$\lim_{n \to \infty} \left(1 + \frac{1}{n} \right) = 1,$$

    $$\lim_{n \to \infty} \left(1 + \frac{1}{n} \right)^3 = \left(\lim_{n \to \infty} \left(1 + \frac{1}{n} \right) \right)^3 = 1^3,$$

    because the exponent, 3, has a fixed value, so we can bring it outside the limit. However, it is *not* true that

    $$\lim_{n \to \infty} \left(1 + \frac{1}{n} \right)^n = \lim_{n \to \infty} 1^n = 1$$

    since the exponent is not fixed, but rather approaching infinity at the same time as the sum $1 + \frac{1}{n}$ is approaching 1. Thus we are applying the limit only to part of the expression, namely the $\frac{1}{n}$, rather than taking the limit of the whole, which gives the incorrect result. In fact we can verify that

    $$2.25 = \left(1 + \frac{1}{2} \right)^2 < \left(1 + \frac{1}{n} \right)^n \quad \text{for any } n > 2.$$

    We will prove this result in Exercise 31 in the Review Exercises for Chapter 4.

    (b) Your calculator first finds $1 + \frac{1}{n}$ and then raises it to the nth power. But it can only do this with a certain number of digits of accuracy, and so if n gets too large, $\frac{1}{n}$ becomes so small that the calculator rounds $1 + \frac{1}{n}$ down to exactly 1, and returns the result $1^n = 1$.

SOLUTIONS TO REVIEW PROBLEMS FOR CHAPTER THREE

1. For $n = 210$, the left sum ≈ 1.466 (rounding up) and the right sum ≈ 1.417 (rounding down). Since the left and right sums differ by 0.049, their average must be within 0.245 of the true value, so the value of the integral will be 1.442 to one decimal place accuracy.

5. For $n = 10$, LEFT ≈ -0.086 (rounding down) and RIGHT ≈ -0.080 (rounding up). Since the left and right sums differ by 0.006, their average must be within 0.003 of the true value, so $\int_2^3 -\frac{1}{(r+1)^2}\, dx = -0.083$ to one decimal place.

9. (a) For $n = 10$, the left sum ≈ 0.054 (rounding down), and the right sum ≈ 0.058 (rounding up). Since the left and right sums differ by 0.004, their average must be within 0.002 of the true value, so the value of the integral is 0.056 to one decimal place.

    (b) Using the Fundamental Theorem of Calculus we find that the integral is $F(0.4) - F(0.2) = \frac{1}{2}(\sin^2(0.4) - \sin^2(0.2)) \approx 0.05609$.

13. (a)

$$\text{Average population} = \frac{1}{10}\int_0^{10} 67.38(1.026)^t\, dt$$

Evaluating the integral numerically gives

$$\text{Average population} \approx 76.8 \text{ million}$$

   (b) In 1980, $t = 0$, and $P = 67.38(1.026)^0 = 67.38$.
   In 1990, $t = 10$, and $P = 67.38(1.026)^{10} = 87.10$.
   Average$= \frac{1}{2}(67.38 + 87.10) = 77.24$ million.

   (c) If P had been linear, the average value found in (a) would have been the one we found in (b). Since the population graph is concave up, it is below the secant line. Thus, the actual values of P are less than the corresponding values on the secant line, and so the average found in (a) is smaller than that in (b).

17. The length of the curve from O to A is surely more than OP + PA. Thinking of OP as the hypotenuse of a right triangle with sides $\frac{\pi}{2}$ and 1, we see that OP $= \sqrt{1 + \frac{\pi^2}{4}} \approx 1.86$. Similarly, PA ≈ 1.86. So the length of OA is greater than 3.72.

Now consider a broken line with six segments as shown. Q $= (\frac{\pi}{6}, \frac{1}{2})$, R $= (\frac{\pi}{3}, \frac{\sqrt{3}}{2})$, S $= (\frac{\pi}{2}, 1)$.

OQ $= \sqrt{(\frac{1}{2})^2 + (\frac{\pi}{6})^2} \approx 0.7240$

QR $= \sqrt{(\frac{1}{2}\sqrt{3} - \frac{1}{2})^2 + (\frac{\pi}{6})^2} \approx 0.6389$

RS $= \sqrt{(1 - \frac{\sqrt{3}}{2})^2 + (\frac{\pi}{6})^2} \approx 0.5405$

The length of the broken line is (using symmetry) $2(OQ + QR + RS) \approx 3.8068$.

The curve must be a little longer than this, but, as is evident from the figure, very little longer, so 3.8 is surely close to the exact length of the curve. We will return to the study of lengths of curves in Chapter 7.

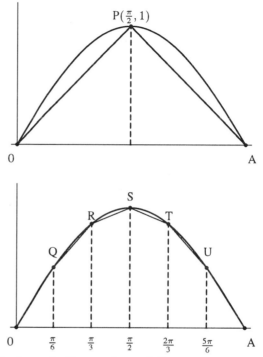

21. (a) The distance traveled is the integral of the velocity, so in T seconds you fall

$$\int_0^T \frac{g}{k}(1 - e^{-kt})\, dt.$$

Putting in the given values we have:

distance fallen in T seconds $= \int_0^T 49(1 - e^{-0.2t})\, dt$ meters.

(b) We want the number T for which

$$\int_0^T 49(1 - e^{-0.2t})\, dt = 5000.$$

Finding T precisely is difficult, but we can show that $T \approx 107$ as follows:

Since your velocity is always less than 49 m/sec it will take more than $\frac{5000}{49} (\approx 102)$ seconds to fall 5000 meters; i.e. $T > 102$.

If we draw the graph of v versus t we see that it has an asymptote $v = 49$. This is called the *terminal velocity*. Terminal velocity is reached, in a practical sense, fairly soon. When $t = 20$, v is already more than 48.1.

When $t = 100$, $v = 48.999\ldots$, so $v \approx 49$ m/sec.

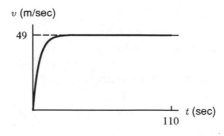

We want to find the time T so that the area under the graph of v from $t = 0$ to $t = T$ is 5000. Since v is increasing, we have:

$$\text{left-hand sum} \leq \int_0^T 49(1 - e^{-0.2t})\,dt \leq \text{right-hand sum}.$$

Since $T > 102$, we try $T = 103$, and using $n = 100$ subdivisions, we have:

$$4775.8 \leq \int_0^{103} 49(1 - e^{-0.2t})\,dt \leq 4826.4$$

Thus, our guess for T is too small, and w need about 173 to 225 meters. since $v \approx 49$ m/sec for $t \geq 1$, we see that each additional second results in approximately 49 meters in distance fallen, so we try $T = 107$. Using $n = 100$ subdivisions, we have:

$$4970.8 \leq \int_0^{107} 49(1 - e^{-0.2t})\,dt \leq 5023.3$$

Therefore $T \approx 107$ seconds.

CHAPTER FOUR

4.1 SOLUTIONS

1. (a) $f(x) = -3x + 2, g(x) = 2x + 1$.

$$
\begin{aligned}
k(x) &= f(x) + g(x) \\
&= (-3x + 2) + (2x + 1) \\
&= -x + 3 \\
k'(x) &= -1.
\end{aligned}
$$

Also, $f'(x) = -3, g'(x) = 2$, so $f'(x) + g'(x) = -3 + 2 = -1$.

(b)

$$
\begin{aligned}
j(x) &= f(x) - g(x) \\
&= (-3x + 2) - (2x + 1) \\
&= -5x + 1 \\
j'(x) &= -5.
\end{aligned}
$$

Also, $f'(x) - g'(x) = -3 - 2 = -5$.

5. Say $\lim_{x \to a} f(x) = A$ and $\lim_{x \to a} g(x) = B$, i.e., as x comes arbitrarily close to a, $f(x)$ comes arbitrarily close to A and $g(x)$ comes arbitrarily close to B. So, as $x \to a$, the sum $f(x) + g(x)$ comes arbitrarily close to $A + B$, i.e.,

$$
\lim_{x \to a} [f(x) + g(x)] = A + B = \lim_{x \to a} f(x) + \lim_{x \to a} g(x).
$$

4.2 SOLUTIONS

1. $y' = 12x^{11}$.

5. $y' = -\frac{3}{4}x^{-\frac{7}{4}}$.

9. $y' = 6x^{\frac{1}{2}} - \frac{5}{2}x^{-\frac{1}{2}}$.

13. $y' = 6t - \frac{6}{t^{3/2}} + \frac{2}{t^3}$.

17. $f(t) = \frac{1}{t^2} + \frac{1}{t} - \frac{1}{t^4} = t^{-2} + t^{-1} - t^{-4}$
$f'(t) = -2t^{-3} - t^{-2} + 4t^{-5}$

21. So far, we can only take the derivative of powers of x and the sums of constant multiples of powers of x. Since we cannot write $\sqrt{x + 3}$ in this form, we cannot yet take its derivative.

25. We cannot write $\frac{1}{3x^2+4}$ as the sum of powers of x multiplied by constants.

29. Once again, the x is in the exponent and we haven't learned how to handle that yet.

33.

$$y' = 3x^2 - 18x - 16 = 5$$
$$3x^2 - 18x - 21 = 0$$
$$x^2 - 6x - 7 = 0$$
$$(x+1)(x-7) = 0$$
$$x = -1 \text{ or } x = 7.$$

$f(-1) = 7, f(7) = -209.$
Thus, the two points are $(-1, 7)$ and $(7, -209)$.

37. (a) $p(x) = x^2 - x$. Now, $p'(x) = 2x - 1 < 0$ when $x < \frac{1}{2}$. So p is decreasing when $x < \frac{1}{2}$.

   (b) $p(x) = x^{\frac{1}{2}} - x$.

$$p'(x) = \frac{1}{2}x^{-\frac{1}{2}} - 1 < 0$$

$$\frac{1}{2}x^{-\frac{1}{2}} < 1$$

$$x^{-\frac{1}{2}} < 2$$

$$x^{\frac{1}{2}} > \frac{1}{2}$$

$$x > \frac{1}{4}.$$

   Thus $p(x)$ is decreasing when $x > \frac{1}{4}$.

   (c) $p(x) = x^{-1} - x$.

$$p'(x) = -1x^{-2} - 1 < 0$$

$$-x^{-2} < 1$$

$$x^{-2} > -1,$$

   which is always true where x^{-2} is defined, since $x^{-2} = \frac{1}{x^2}$ is always positive. Thus $p(x)$ is always decreasing, unless $x = 0$.

41. $\dfrac{dF}{dr} = -\dfrac{2GMm}{r^3}.$

44. (a) $A = \pi r^2$
       $\frac{dA}{dr} = 2\pi r.$

   (b) This is the formula for the circumference of a circle.

(c) $A'(r) = \lim\limits_{h \to 0} \dfrac{A(r+h)-A(r)}{h}$

The numerator of the difference quotient denotes the area contained between the inner circle (radius r) and the outer circle (radius $r+h$). As h approaches 0, this area can be approximated by the product of the circumference of the inner circle and the "width" of the area, i.e., h. Dividing this by the denominator, h, we get $A' =$ the circumference of the circle with radius r.

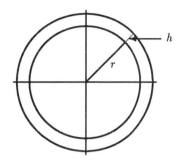

We can also think about the derivative of A as the rate of change of area for a small change in radius. If the radius increases by a tiny amount, the area will increase by a thin ring whose area is simply the circumference at that radius times the small amount. To get the rate of change, we divide by the small amount and obtain the circumference.

45. $V = \frac{4}{3}\pi r^3$

$\frac{dV}{dr} = 4\pi r^2 =$ surface area of a sphere.

Our reasoning is similar to that of Problem 44. The difference quotient $\frac{V(r+h)-V(r)}{h}$ is the volume between two spheres divided by the change in radius. Furthermore, when h is very small (and consequently $V(r+h) \approx V(r)$) this volume is like a coating of paint of depth h applied to the surface of the sphere. The volume of the paint is about $h \cdot$ (Surface Area) for small h: dividing by h gives back the surface area. Also, thinking about the derivative as the rate of change of the function for a small change in the variable, the answer seems clear. If you increase the radius of a sphere the tiniest amount, the volume will increase by a very thin layer whose volume will be the surface area at that radius multiplied by that tiniest amount.

4.3 SOLUTIONS

1. $y' = 10t + 4e^t$.

5. $\dfrac{dy}{dx} = 3 - 2(\ln 4)4^x$.

9. $f(t) = e^t \cdot e^2$. Then, since e^2 is just a constant, $f'(t) = \frac{d}{dt}(e^t e^2) = e^2 \frac{d}{dt}e^t = e^2 e^t = e^{t+2}$.

13. $f'(z) = (2\ln 3)z + (\ln 4)e^z$.

17. $\dfrac{dy}{dx} = \pi^x \ln \pi$

21.   We can take the derivative of the sum $x^2 + 2^x$, but not the product.

25. $f(s) = 5^s e^s = (5e)^s$, so $f'(s) = \ln(5e) \cdot (5e)^s = (1 + \ln 5)5^s e^s$.

29.   This is the composition of two functions each of which we can take the derivative of, but we don't know how to take the derivative of the composition.

33.   Since $P = 1 \cdot (1.05)^t$, $\frac{dP}{dt} = \ln(1.05)1.05^t$. When $t = 10$,

$$\frac{dP}{dt} = (\ln 1.05)(1.05)^{10} \approx \$0.07947/\text{year} \approx 7.95¢/\text{year}.$$

37.   We are interested in when the derivative $\dfrac{d(a^x)}{dx}$ is positive and when it is negative. The quantity a^x is always positive. However $\ln a > 0$ for $a > 1$ and $\ln a < 0$ for $0 < a < 1$. Thus the function a^x is increasing for $a > 1$ and decreasing for $a < 1$.

4.4 SOLUTIONS

1.   By the product rule,

$$f'(x) = 2x(x^3 + 5) + x^2(3x^2) = 2x^4 + 3x^4 + 10x = 5x^4 + 10x.$$

Alternatively,

$$f'(x) = (x^5 + 5x^2)' = 5x^4 + 10x.$$

The two answers should, and do, match.

5. $y' = 2^x + x(\ln 2)2^x = 2^x(1 + x \ln 2)$.

9.   It is easier to do this by multipling it out first, rather than using the product rule first:
$$z = s^4 - s, \quad z' = 4s^3 - 1.$$

13. $z' = \dfrac{(2t + 5)(t + 3) - (t^2 + 5t + 2)}{(t + 3)^2} = \dfrac{t^2 + 6t + 13}{(t + 3)^2}$.

17.
$$\begin{aligned}
f'(x) &= \frac{(2 + 3x + 4x^2)(1) - (1 + x)(3 + 8x)}{(2 + 3x + 4x^2)^2} \\
&= \frac{2 + 3x + 4x^2 - 3 - 11x - 8x^2}{(2 + 3x + 4x^2)^2} \\
&= \frac{-4x^2 - 8x - 1}{(2 + 3x + 4x^2)^2}.
\end{aligned}$$

18.   Notice that you can cancel a z out of the numerator and denominator to get

$$f(z) = \frac{3z}{5z + 7}, \qquad z \neq 0$$

Then

$$f'(z) = \frac{(5z+7)3 - 3z(5)}{(5z+7)^2}$$

$$= \frac{15z + 21 - 15z}{(5z+7)^2}$$

$$= \frac{21}{(5z+7)^2}, z \neq 0.$$

[If you used the quotient rule correctly without canceling the z out first, your answer should simplify to this one, but it is usually a good idea to simplify as much as possible before differentiating.]

21. $f(x) = e^x \cdot e^x$
$f'(x) = e^x \cdot e^x + e^x \cdot e^x = 2e^{2x}.$

25. This is the same function we were asked to differentiate in Problem 18, so we know that, if $x \neq 0$,

$$f'(x) = \frac{21}{(5x+7)^2}.$$

So at $x = 1$,

$$y = f(1) = \frac{3}{12} = \frac{1}{4},$$

$$y' = \frac{21}{144} = \frac{7}{48}.$$

So,

$$y - \frac{1}{4} = \frac{7}{48}(x - 1).$$

$$y = \frac{7}{48}x + \frac{5}{48}.$$

29. (a) $f'(x) = (x - 2) + (x - 1).$
(b) Think of f as the product of two factors, with the first as $(x - 1)(x - 2)$. (The reason for this is that we have already differentiated $(x - 1)(x - 2)$).

$$f(x) = [(x - 1)(x - 2)](x - 3).$$

Now $f'(x) = [(x - 1)(x - 2)]'(x - 3) + [(x - 1)(x - 2)](x - 3)'$
Using the result of a):

$$f'(x) = [(x - 2) + (x - 1)](x - 3) + [(x - 1)(x - 2)] \cdot 1$$
$$= (x - 2)(x - 3) + (x - 1)(x - 3) + (x - 1)(x - 2).$$

(c) Because we have already differentiated $(x-1)(x-2)(x-3)$, rewrite f as the product of two factors, the first being $(x-1)(x-2)(x-3)$:

$$f(x) = [(x-1)(x-2)(x-3)](x-4)$$

Now $f'(x) = [(x-1)(x-2)(x-3)]'(x-4) + [(x-1)(x-2)(x-3)](x-4)'$.

$$\begin{aligned} f'(x) &= [(x-2)(x-3) + (x-1)(x-3) + (x-1)(x-2)](x-4) \\ &\quad + [(x-1)(x-2)(x-3)] \cdot 1 \\ &= (x-2)(x-3)(x-4) + (x-1)(x-3)(x-4) \\ &\quad + (x-1)(x-2)(x-4) + (x-1)(x-2)(x-3). \end{aligned}$$

From the solutions above, one can observe that when f is a product, its derivative is obtained by differentiating each factor in turn (leaving the other factors alone), and adding the results.

33. (a) $f(140) = 15,000$ says that 15,000 skateboards are sold when the cost is \$140 per board. $f'(140) = -100$ means that if the price is increased from \$140, roughly speaking, every dollar of increase will decrease the total sales by 100 boards.

(b) $$\frac{dR}{dp} = \frac{d}{dp}(p \cdot q) = \frac{d}{dp}(p \cdot f(p)) = f(p) + pf'(p).$$

So,

$$\left. \frac{dR}{dp} \right|_{p=140} = f(140) + 140f'(140)$$

$$= 15,000 + 140(-100) = 1000.$$

(c) From (b) we see that $\left. \dfrac{dR}{dp} \right|_{p=140} = 1000 > 0$. This means that the revenue will increase if the price is raised. (Note we can only be certain that the revenue will increase for a relatively small price increase.)

4.5 SOLUTIONS

1. $f'(x) = 99(x+1)^{98} \cdot 1 = 99(x+1)^{98}$.

5. $w' = 100(\sqrt{t} + 1)^{99} \left(\frac{1}{2\sqrt{t}} \right) = \frac{50}{\sqrt{t}}(\sqrt{t} + 1)^{99}$.

9. $y' = \dfrac{3s^2}{2\sqrt{s^3 + 1}}$.

12. $f'(z) = \dfrac{1}{2\sqrt{z}} e^{-z} - \sqrt{z} e^{-z}$.

13. We can write this as $f(z) = \sqrt{z}e^{-z}$, in which case it is the same as problem 12. So $f'(z) = \dfrac{1}{2\sqrt{z}} e^{-z} - \sqrt{z} e^{-z}$.

17. $f'(\theta) = -1(1 + e^{-\theta})^{-2}(e^{-\theta})(-1) = \dfrac{e^{-\theta}}{(1 + e^{-\theta})^2}.$

21.

$$f(y) = \left[10^{(5-y)}\right]^{\frac{1}{2}} = 10^{\frac{5}{2} - \frac{1}{2}y}$$

$$f'(y) = (\ln 10)\left(10^{\frac{5}{2} - \frac{1}{2}y}\right)\left(-\frac{1}{2}\right) = -\frac{1}{2}(\ln 10)(10^{\frac{5}{2} - \frac{1}{2}y}).$$

25.

$$f(x) = 6e^{5x} + e^{-x^2} \qquad\qquad f'(x) = 30e^{5x} - 2xe^{-x^2}$$
$$f(1) = 6e^5 + e^{-1} \qquad\qquad f'(1) = 30e^5 - 2(1)e^{-1}$$

$$y - y_1 = m(x - x_1)$$
$$y - (6e^5 + e^{-1}) = (30e^5 - 2e^{-1})(x - 1)$$
$$y - (6e^5 + e^{-1}) = (30e^5 - 2e^{-1})x - (30e^5 - 2e^{-1})$$
$$y = (30e^5 - 2e^{-1})x - 30e^5 + 2e^{-1} + 6e^5 + e^{-1}$$
$$\approx 4451.66x - 3560.81.$$

29.   Yes. To see why, simply plug $x = \sqrt[3]{2t + 5}$ into the expression $3x^2 \dfrac{dx}{dt}$ and evaluate it. To do this, first we calculate $\dfrac{dx}{dt}$. By the chain rule,

$$\frac{dx}{dt} = \frac{d}{dt}(2t + 5)^{\frac{1}{3}} = \frac{2}{3}(2t + 5)^{-\frac{2}{3}} = \frac{2}{3}[(2t + 5)^{\frac{1}{3}}]^{-2}.$$

But since $x = (2t + 5)^{\frac{1}{3}}$, we have (by substitution)

$$\frac{dx}{dt} = \frac{2}{3}x^{-2}.$$

It follows that $3x^2 \dfrac{dx}{dt} = 3x^2 \left(\dfrac{2}{3}x^{-2}\right) = 2.$

33.   (a)

$$\frac{dm}{dv} = \frac{d}{dv}\left[m_0\left(1 - \frac{v^2}{c^2}\right)^{-1/2}\right]$$

$$= m_0\left(-\frac{1}{2}\right)\left(1 - \frac{v^2}{c^2}\right)^{-3/2}\left(-\frac{2v}{c^2}\right)$$

$$= \frac{m_0 v}{c^2}\frac{1}{\sqrt{\left(1 - \frac{v^2}{c^2}\right)^3}}.$$

(b) $\dfrac{dm}{dv}$ represents the rate of change of mass with respect to the speed v.

4.6 SOLUTIONS

1.
$$f(x) = (1 - \cos x)^{\frac{1}{2}}$$
$$f'(x) = \frac{1}{2}(1 - \cos x)^{-\frac{1}{2}}(-(-\sin x))$$
$$= \frac{\sin x}{2\sqrt{1 - \cos x}}.$$

5. $w' = e^t \cos(e^t)$.

9. $z' = e^{\cos \theta} - \theta(\sin \theta)e^{\cos \theta}$.

13.
$$f'(x) = (e^{-2x})(-2)(\sin x) + (e^{-2x})(\cos x)$$
$$= -2\sin x(e^{-2x}) + (e^{-2x})(\cos x)$$
$$= e^{-2x}[\cos x - 2\sin x].$$

17. $z' = \dfrac{-3e^{-3\theta}}{\cos^2(e^{-3\theta})}$.

21.

TABLE 4.1

x	$\cos x$	Difference Quotient	$-\sin x$
0	1.0	−0.0005	0.0
0.1	0.995	−0.10033	−0.099833
0.2	0.98007	−0.19916	−0.19867
0.3	0.95534	−0.296	−0.29552
0.4	0.92106	−0.38988	−0.38942
0.5	0.87758	−0.47986	−0.47943
0.6	0.82534	−0.56506	−0.56464

25. f is certainly periodic. Since sine is periodic with period 2π, we know that $\sin w = \sin(w + 2\pi)$. Since these have the same value, $e^{\sin w} = e^{\sin(w+2\pi)}$, so f also has period 2π. It does not grow unboundedly, though. Sine cannot get bigger than 1 or less than -1. Consequently, the biggest f can get is $e^1 = e$ and the smallest is $e^{-1} = \frac{1}{e}$. (f has no zeros and is never negative.) See the following figure.

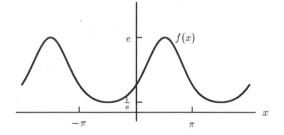

29.

$$w'(t) = u'(v(t)) \cdot v'(t) = u'(\cos t)(-\sin t)$$
$$w'(\pi) = u'(\cos \pi)(-\sin \pi)$$
$$= u'(-1)(-\sin \pi)$$
$$= 2 \cdot 0$$
$$= 0.$$

33. (a) If $f(x) = \sin x$, then

$$
\begin{aligned}
f'(x) &= \lim_{h \to 0} \frac{\sin(x + h) - \sin x}{h} \\
&= \lim_{h \to 0} \frac{(\sin x \cos h + \sin h \cos x) - \sin x}{h} \\
&= \lim_{h \to 0} \frac{\sin x(\cos h - 1) + \sin h \cos x}{h} \\
&= \sin x \lim_{h \to 0} \frac{\cos h - 1}{h} + \cos x \lim_{h \to 0} \frac{\sin h}{h}.
\end{aligned}
$$

(b) $\frac{\cos h - 1}{h} \to 0$ and $\frac{\sin h}{h} \to 1$, as $h \to 0$. Thus, $f'(x) = \sin x \cdot 0 + \cos x \cdot 1 = \cos x$.

(c) Similarly,

$$
\begin{aligned}
g'(x) &= \lim_{h \to 0} \frac{\cos(x + h) - \cos x}{h} \\
&= \lim_{h \to 0} \frac{(\cos x \cos h - \sin x \sin h) - \cos x}{h} \\
&= \lim_{h \to 0} \frac{\cos x(\cos h - 1) - \sin x \sin h}{h} \\
&= \cos x \lim_{h \to 0} \frac{\cos h - 1}{h} - \sin x \lim_{h \to 0} \frac{\sin h}{h} \\
&= -\sin x.
\end{aligned}
$$

4.7 SOLUTIONS

1. $f'(x) = \frac{-1}{1-x} = \frac{1}{x-1}$.

5. $f'(x) = \frac{1}{1-e^{-x}} \cdot -e^{-x}(-1) = \frac{e^{-x}}{1-e^{-x}}$.

9. $f'(x) = \frac{1}{e^{7x}} \cdot (e^{7x})7 = 7$.
   (Note also that $\ln(e^{7x}) = 7x$ implies $f'(x) = 7$.)

13. $f'(y) = \dfrac{2y}{\sqrt{1-y^4}}$.

17. Let

$$g(x) = \arcsin x$$

so

$$\sin[g(x)] = x.$$

Differentiating,

$$\cos[g(x)] \cdot g'(x) = 1$$

$$g'(x) = \frac{1}{\cos[g(x)]}$$

Using the fact that $\sin^2 \theta + \cos^2 \theta = 1$, and $\cos[g(x)] \geq 0$, since $-\frac{\pi}{2} \leq g(x) \leq \frac{\pi}{2}$, we get

$$\cos[g(x)] = \sqrt{1 - (\sin[g(x)])^2}.$$

Therefore,

$$g'(x) = \frac{1}{\sqrt{1 - (\sin[g(x)])^2}}$$

Since $\sin[g(x)] = x$, we have

$$g'(x) = \frac{1}{\sqrt{1 - x^2}}, -1 < x < 1.$$

21. The closer you look at the function, the more it begins to look like a line with slope equal to the derivative of the function at $x = 0$. Hence, functions whose derivatives at $x = 0$ are equal will look the same there.

   The following functions look like the line $y = x$ since, in all cases, $y' = 1$ at $x = 0$.
   $y = x$ $y' = 1$
   $y = \sin x$ $y' = \cos x$
   $y = \tan x$ $y' = \frac{1}{\cos^2 x}$
   $y = \ln(x+1)$ $y' = \frac{1}{x+1}$

The following functions look like the line $y = 0$ since, in all cases, $y' = 0$ at $x = 0$.

$y = x^2$ $y' = 2x$

$y = x \sin x$ $y' = x \cos x + \sin x$

$y = x^3$ $y' = 3x^2$

$y = \frac{1}{2} \ln (x^2 + 1)$ $y' = 2x \cdot \frac{1}{2} \cdot \frac{1}{x^2+1} = \frac{x}{x^2+1}$

$y = 1 - \cos x$ $y' = \sin x$

The following functions look like the line $x = 0$ since, in all cases, as $x \to 0^+$, the slope $y' \to \infty$.

$y = \sqrt{x}$ $y' = \frac{1}{2\sqrt{x}}$

$y = \sqrt{\frac{x}{x+1}}$ $y' = \frac{(x+1)-x}{(x+1)^2} \cdot \frac{1}{2} \cdot \frac{1}{\sqrt{\frac{x}{x+1}}} = \frac{1}{2(x+1)^2} \cdot \sqrt{\frac{x+1}{x}}$

$y = \sqrt{2x - x^2}$ $y' = (2 - 2x)\frac{1}{2} \cdot \frac{1}{\sqrt{2x-x^2}} = \frac{1-x}{\sqrt{2x-x^2}}$

25. (a) Since the elevator is descending at 30 ft/sec, its height from the ground is given by $h(t) = 300 - 30t$, for $0 \le t \le 10$.

(b) From the triangle in the figure,

$$\tan \theta = \frac{h(t) - 100}{150} = \frac{300 - 30t - 100}{150} = \frac{200 - 30t}{150}.$$

Therefore

$$\theta = \arctan \left(\frac{200 - 30t}{150} \right)$$

and

$$\frac{d\theta}{dt} = \frac{1}{1 + \left(\frac{200-30t}{150}\right)^2} \cdot \left(\frac{-30}{150}\right) = -\frac{1}{5} \left(\frac{150^2}{150^2 + (200 - 30t)^2} \right).$$

Notice that $\frac{d\theta}{dt}$ is always negative, which is reasonable since θ decreases as the elevator descends.

(c) If we want to know when θ changes (decreases) the fastest, we want to find out when $d\theta/dt$ has the largest magnitude. This will occur when the denominator, $150^2 + (200 - 30t)^2$, in the expression for $d\theta/dt$ is the smallest, or when $200 - 30t = 0$. This occurs when $t = \frac{200}{30}$ seconds, and so $h(\frac{200}{30}) = 100$ feet, i.e., when the elevator is at the level of the observer.

4.8 SOLUTIONS

1. We differentiate implicitly both sides of the equation with respect to x.

$$2x + \left(y + x\frac{dy}{dx} \right) - 3y^2\frac{dy}{dx} = y^2 + x(2y)\frac{dy}{dx} ,$$

$$x\frac{dy}{dx} - 3y^2\frac{dy}{dx} - 2xy\frac{dy}{dx} = y^2 - y - 2x ,$$

$$\frac{dy}{dx} = \frac{y^2 - y - 2x}{x - 3y^2 - 2xy}.$$

5. We differentiate implicitly both sides of the equation with respect to x.

$$\cos(xy)\left(y + x\frac{dy}{dx}\right) = 2\,,$$

$$y\cos(xy) + x\cos(xy)\frac{dy}{dx} = 2\,,$$

$$\frac{dy}{dx} = \frac{2 - y\cos(xy)}{x\cos(xy)}.$$

9. First, we must find the slope of the tangent, i.e. $\left.\dfrac{dy}{dx}\right|_{(1,-1)}$. Differentiating implicitly, we have:

$$y^2 + x(2y)\frac{dy}{dx} = 0,$$

$$\frac{dy}{dx} = -\frac{y^2}{2xy} = -\frac{y}{2x}.$$

Substitution yields $\left.\dfrac{dy}{dx}\right|_{(1,-1)} = -\dfrac{-1}{2} = \dfrac{1}{2}$. Using the point-slope formula for a line, we have that the equation for the tangent line is $y + 1 = \frac{1}{2}(x - 1)$ or $y = \frac{1}{2}x - \frac{3}{2}$.

13. $y = x^{\frac{m}{n}}$. Taking n^{th} powers of both sides of this expression yields $(y)^n = (x^{\frac{m}{n}})^n$, or $y^n = x^m$.

$$\frac{d}{dx}(y^n) = \frac{d}{dx}(x^m)$$

$$ny^{n-1}\frac{dy}{dx} = mx^{m-1}$$

$$\frac{dy}{dx} = \frac{m}{n}\frac{x^{m-1}}{y^{n-1}}$$

$$= \frac{m}{n}\frac{x^{m-1}}{(x^{m/n})^{n-1}}$$

$$= \frac{m}{n}\frac{x^{m-1}}{x^{m-\frac{m}{n}}}$$

$$= \frac{m}{n}x^{(m-1)-(m-\frac{m}{n})}$$

$$= \frac{m}{n}x^{\frac{m}{n}-1}.$$

17. (a) Solving for $\frac{dy}{dx}$ by implicit differentiation yields

$$3x^2 + 3y^2\frac{dy}{dx} - y^2 - 2xy\frac{dy}{dx} = 0$$
$$\frac{dy}{dx} = \frac{y^2 - 3x^2}{3y^2 - 2xy}.$$

(b) We can approximate the curve near $x = 1$, $y = 2$ by its tangent line. The tangent line will have slope $\frac{(2)^2 - 3(1)^2}{3(2)^2 - 2(1)(2)} = \frac{1}{8} = 0.125$. Thus its equation is

$$y = 0.125x + 1.875$$

Using the y values of the tangent line to approximate the y values of the curve, we get:

TABLE 4.2

x	0.96	0.98	1	1.02	1.04
approximate y	1.995	1.9975	2.000	2.0025	2.005

(c) When $x = 0.96$, we get the equation $0.96^3 + y^3 - 0.96y^2 = 5$, whose solution by numerical methods is 1.9945, which is close to the one above.

(d) The tangent line is horizontal when $\frac{dy}{dx}$ is zero and vertical when $\frac{dy}{dx}$ is undefined. These will occur when the numerator is zero and when the denominator is zero, respectively.

Thus, we know that the tangent is horizontal when $y^2 - 3x^2 = 0 \Rightarrow y = \pm\sqrt{3}x$. To find the points that satisfy this condition, we substitute back into the original equation for the curve:

$$x^3 + y^3 - xy^2 = 5$$
$$x^3 \pm 3\sqrt{3}x^3 - 3x^3 = 5$$
$$x^3 = \frac{5}{\pm 3\sqrt{3} - 2}$$
So $x \approx 1.1609$ or $x \approx -0.8857$.

Substituting,

$$y = \pm\sqrt{3}x \quad \text{so} \quad y \approx 2.0107 \quad \text{or} \quad y \approx 1.5341.$$

Thus, the tangent line is horizontal at $(1.1609, 2.0107)$ and $(-0.8857, 1.5341)$.

Also, we know that the tangent is vertical whenever $3y^2 - 2xy = 0$, that is, when $y = \frac{2}{3}x$ or $y = 0$. Substituting into the original equation for the curve gives us $x^3 + (\frac{2}{3}x)^3 - (\frac{2}{3})^2 x^3 = 5$, whence $x^3 \approx 5.8696$, so $x \approx 1.8039$, $y \approx 1.2026$. The other vertical tangent is at $y = 0$, $x = \sqrt[3]{5}$.

4.9 SOLUTIONS

1. With $f(x) = 1/x$, we see that the tangent line approximation to f near $x = 1$ is

$$f(x) \approx f(1) + f'(1)(x - 1),$$

which becomes

$$\frac{1}{x} \approx 1 + f'(1)(x - 1).$$

Since $f'(x) = -1/x^2$, $f'(1) = -1$. Thus our formula reduces to

$$\frac{1}{x} \approx 1 - (x - 1) = 2 - x.$$

This is the local linearization of $1/x$ near $x = 1$.

5. If $\$P$ were deposited, then Pe^{rt} would be the balance after t years if interest were compounded continuously at the nominal rate r (see Section 4.4). If interest were compounded n times a year, then the balance would be $P(1 + \frac{r}{n})^{nt}$. The local linearization $e^{rt} \approx 1 + rt$ tells us that for small values of t, say $t = \frac{1}{n}$,

$$Pe^{r(\frac{1}{n})} \approx P\left(1 + r\left(\frac{1}{n}\right)\right) = P(1 + \frac{1}{n}).$$

In other words, the balance after one compounding period is approximately the same whether interest is compounded n times a year or continuously.

9. (a) Let $f(x) = 1/(1+x)$. Then $f'(x) = -1/(1+x)^2$ by the chain rule. So $f(0) = 1$, and $f'(0) = -1$. Therefore, for x near 0, $1/(1 + x) \approx f(0) + f'(0)x = 1 - x$.

   (b) We know that for small y, $1/(1 + y) \approx 1 - y$. Let $y = x^2$; when x is small, so is $y = x^2$. Hence, for small x, $1/(1 + x^2) \approx 1 - x^2$.

   (c) Since the linearization of $1/(1 + x^2)$ is the line $y = 1$, and this line has a slope of 0, the derivative of $1/(1 + x^2)$ is zero at $x = 0$.

SOLUTIONS TO REVIEW PROBLEMS FOR CHAPTER FOUR

1. $f'(x) =$
   $6x(e^x - 4) + (3x^2 + \pi)e^x = 6xe^x - 24x + 3x^2 e^x + \pi e^x.$

5. $\dfrac{d}{dt} e^{(1+3t)^2} = e^{(1+3t)^2} \cdot 2(1 + 3t) \cdot 3 = 6(1 + 3t)e^{(1+3t)^2}.$

9. $\dfrac{d}{dy} \ln \ln(2y^3) = \dfrac{1}{\ln(2y^3)} \dfrac{1}{2y^3} 6y^2.$

13. $\dfrac{d}{d\theta}\sqrt{a^2 - \sin^2\theta} = \dfrac{1}{2\sqrt{a^2 - \sin^2\theta}}(-2\sin\theta\cos\theta) = -\dfrac{\sin\theta\cos\theta}{\sqrt{a^2 - \sin^2\theta}}.$

17. $h'(t) = \dfrac{d}{dt}\left(\ln\left(e^{-t} - t\right)\right) = \dfrac{1}{e^{-t} - t}\left(-e^{-t} - 1\right).$

21. $\dfrac{d}{dy}\left(\dfrac{y}{\cos y + a}\right) = \dfrac{\cos y + a - y(-\sin y)}{(\cos y + a)^2} = \dfrac{\cos y + a + y\sin y}{(\cos y + a)^2}.$

25. (a) $\dfrac{dg}{dr} = GM\dfrac{d}{dr}\left(\dfrac{1}{r^2}\right) = GM\dfrac{d}{dr}\left(r^{-2}\right) = GM(-2)r^{-3} = -\dfrac{2GM}{r^3}.$

(b) $\dfrac{dg}{dr}$ is the rate of change of acceleration due to the pull of gravity. The further away from the center

of the earth, the weaker the pull of gravity is. So g is decreasing and therefore its derivative, $\dfrac{dg}{dr}$,

is negative.

(c) By part (a),

$$\begin{aligned}
\left.\dfrac{dg}{dr}\right|_{r=6400} &= -\left.\dfrac{2GM}{r^3}\right|_{r=6400} \\
&= -\dfrac{2(6.67 \times 10^{-20})(6 \times 10^{24})}{(6400)^3} \\
&\approx -3.05 \times 10^{-6}.
\end{aligned}$$

(d) It is reasonable to assume that g is a constant near the surface of the earth.

29.

$$\dfrac{dy}{dt} = -7.5(0.507)\sin(0.507t) = -3.80\sin(0.507t)$$

(a) When $t = 6$, $\dfrac{dy}{dt} = -3.80\sin(0.507\cdot 6) = -0.38$ meters/hour. So it's falling at 0.38 meters/hour.

(b) When $t = 9$, $\dfrac{dy}{dt} = -3.80\sin(0.507\cdot 9) = 3.76$ meters/hour. So it's rising at 3.76 meters/hour.

(c) When $t = 12$, $\dfrac{dy}{dt} = -3.80\sin(0.507\cdot 12) = 0.75$ meters/hour. So it's rising at 0.75 meters/hour.

(d) When $t = 18$, $\dfrac{dy}{dt} = -3.80\sin(0.507\cdot 18) = -1.12$ meters/hour. So it's falling at 1.12 meters/hour.

33. Since we're given the instantaneous rate of change T at $t = 30$ is 2, we want to choose a and b so that the derivative of T agrees with this value. Differentiating, $T'(t) = ab \cdot c^{-bt}$. Then we have

$$2 = T'(30) = abe^{-30b} \text{ or } e^{-30b} = \dfrac{2}{ab}$$

We also know that at $t = 30$, $T = 120$, so

$$120 = T(30) = 200 - ae^{-30b} \text{ or } e^{-30b} = \dfrac{80}{a}$$

Thus $\dfrac{80}{a} = e^{-30b} = \dfrac{2}{ab}$, so $b = \frac{1}{40} = 0.025$ and $a = 169.36$.

37.   (a)   We can approximate $\frac{d}{dx}[F(x)G(x)H(x)]$ using the large rectangular solids by which our original cube is increased:

$$\text{Volume of whole } - \text{ volume of original solid } = \text{ change in volume.}$$

$$F(x+h)G(x+h)H(x+h) - F(x)G(x)H(x) = \text{change in volume.}$$

The volume of this slab is $F'(x)G(x)H(x)h$

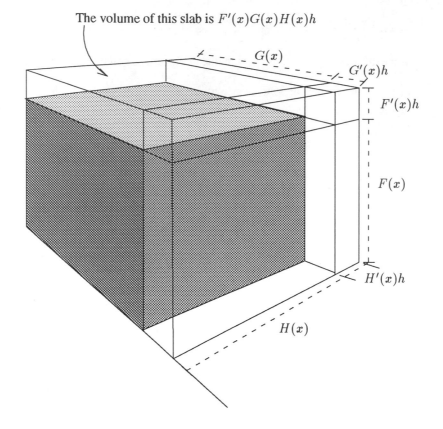

As in the book, we will ignore the <u>smaller</u> regions which are added (the long, thin rectangular boxes and the small cube in the corner.) This can be justified by recognizing that as $h \to 0$, these volumes will shrink much faster than the volumes of the big slabs and will therefore be insignificant. (Note that these smaller regions have an h^2 or h^3 in the formulas of their volumes.) Then we can approximate the change in volume above by:

$$F(x+h)G(x+h)H(x+h) - F(x)G(x)H(x)$$
$$\approx F'(x)G(x)H(x)h \quad \text{(top slab)}$$
$$+ F(x)G'(x)H(x)h \quad \text{(front slab)}$$
$$+ F(x)G(x)H'(x)h \quad \text{(other slab)},$$

dividing by h gives

$$\frac{F(x+h)G(x+h)H(x+h) - F(x)G(x)H(x)}{h}$$
$$\approx F'(x)G(x)H(x) + F(x)G'(x)H(x) + F(x)G(x)H'(x),$$

letting $h \to 0$

$$(FGH)' = F'GH + FG'H + FGH'.$$

(b) Verifying,

$$\frac{d}{dx}[(F(x) \cdot G(x)) \cdot H(x)]$$
$$= (F \cdot G)'(H) + (F \cdot G)(H)'$$
$$= [F'G + FG']H + FGH'$$
$$= F'GH + FG'H + FGH'$$

as before.

(c) From the answer to (b), we observe that the derivative of a product is obtained by differentiating each factor in turn (leaving the other factors alone), and adding the results. So, in general,

$$(f_1 \cdot f_2 \cdot f_3 \cdot \ldots \cdot f_n)' = f_1' f_2 f_3 \cdots f_n + f_1 f_2' f_3 \cdots f_n + \cdots + f_1 \cdots f_{n-1} f_n'.$$

CHAPTER FIVE

5.1 SOLUTIONS

1.

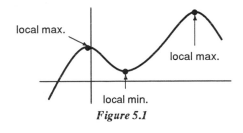

local max.

local max.

local min.

Figure 5.1

5. $f'(x) = 6x^2 + 6x - 36$. To find critical points, we set $f'(x) = 0$. Then

$$6(x^2 + x - 6) = 6(x + 3)(x - 2) = 0.$$

Therefore, the critical points of f are $x = -3$ and $x = 2$. To the left of $x = -3$, $f'(x) > 0$. Between $x = -3$ and $x = 2$, $f'(x) < 0$. To the right of $x = 2$, $f'(x) > 0$. Thus $f(-3)$ is a local maximum, $f(2)$ a local minimum.

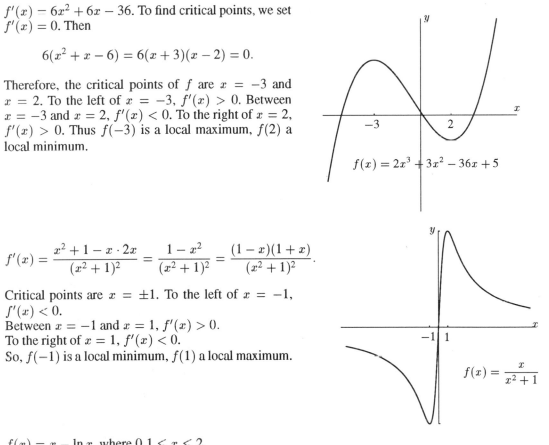

$$f(x) = 2x^3 + 3x^2 - 36x + 5$$

9.

$$f'(x) = \frac{x^2 + 1 - x \cdot 2x}{(x^2 + 1)^2} = \frac{1 - x^2}{(x^2 + 1)^2} = \frac{(1 - x)(1 + x)}{(x^2 + 1)^2}.$$

Critical points are $x = \pm 1$. To the left of $x = -1$, $f'(x) < 0$.
Between $x = -1$ and $x = 1$, $f'(x) > 0$.
To the right of $x = 1$, $f'(x) < 0$.
So, $f(-1)$ is a local minimum, $f(1)$ a local maximum.

$$f(x) = \frac{x}{x^2 + 1}$$

13. $f(x) = x - \ln x$, where $0.1 \le x \le 2$.

(a) $f'(x) = 1 - \frac{1}{x}$. This is zero only when $x = 1$, and so $x = 1$ is the only critical point of f. $f'(x)$ is positive when $x > 1$, and negative when $x < 1$. Thus $f(1) = 1$ is a local minimum.

(b)   We have, by looking at the endpoints and the critical point,

$$f(0.1) = 0.1 - \ln(0.1) \approx 2.4026$$
$$f(1) = 1$$
$$f(2) = 2 - \ln 2 \approx 1.3069.$$

Thus $x = 0.1$ gives the global maximum and $x = 1$ gives the global minimum.

17.

$$f(x) = \frac{x + 50}{x^2 + 525},$$

$$f'(x) = \frac{x^2 + 525 - 2x(x + 50)}{(x^2 + 525)^2} = \frac{-x^2 - 100x + 525}{(x^2 + 525)^2}.$$

Since the denominator is positive for all x, setting $f'(x) = 0$ gives:

$$x^2 + 100x - 525 = 0$$

$$(x - 5)(x + 105) = 0$$

$$x = 5, -105$$

These are the only two critical points. Checking signs using the formula for $f'(x)$, we find $f(x)$ is decreasing for $-105 < x < 5$ and $f(x)$ is increasing for $x < -105$ and $x > 5$.

It is difficult to solve this problem with a calculator for two reasons. First, the critical points are at -105 and 5, which are very far apart. To find both of these points, one would have to search a large portion of the domain. Second, the range of values around the critical points is very small. For example,

$$f(-100) = -0.0047506$$
$$f(-105) = -0.0047619$$
$$f(-110) = -0.0047525$$

As you can see, although $x = -105$ gives a local minimum, f is extremely flat in the neighborhood of -105. Even 5 units away, the difference is less than 1% of the value at -105. It is very hard to see these small differences on a calculator, so solving the problem on a calculator is difficult.

21.

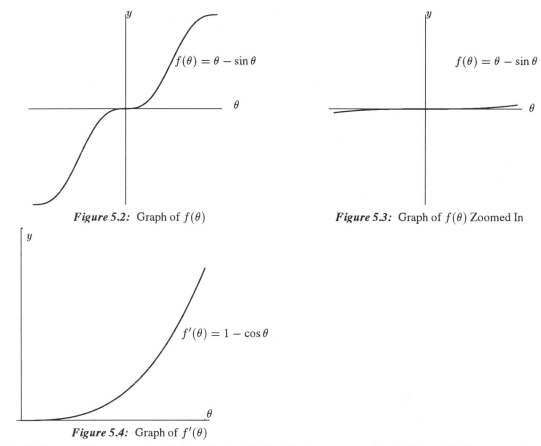

Figure 5.2: Graph of $f(\theta)$

Figure 5.3: Graph of $f(\theta)$ Zoomed In

Figure 5.4: Graph of $f'(\theta)$

(a) In Figure 5.2, we see that $f(\theta) = \theta - \sin\theta$ definitely has a zero at $\theta = 0$. To see if it has any other zeros near the origin, we use our calculator to zoom in. (See Figure 5.3.) No extra root seems to appear no matter how close to the origin we zoom. However, zooming can never tell you for sure that there is not a root that you have not found yet.

(b) Using the derivative we can argue that for sure that there is no other zero. $f'(\theta) - 1 - \cos\theta$. Since $\cos\theta < 1$ for $0 < \theta \le 1$, $f'(\theta) > 0$ for $0 < \theta \le 1$. Thus, f increases for $0 < \theta \le 1$. Consequently, we conclude that the only zero of f is the one at the origin. If f had another zero at x_0, $x_0 > 0$, f would have to "turn around", and recross the x-axis at x_0. But if this were the case, f' would be nonpositive somewhere, which we know to be impossible.

25. (a)

$$P(t) = \frac{2000}{1 + e^{(5.3 - 0.4t)}}$$

Population of Rabbits

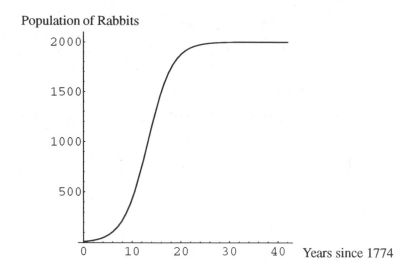

(b)   The population appears to have been growing fastest when there were about 500 rabbits, just over 10 years after Captain Cook left them there.

(c)   The rabbits reproduce quickly, so their population initially grew very rapidly. Limited food and space availability or perhaps predators on the island probably accounts for the population being unable to grow past 2000.

5.2   SOLUTIONS

1.   To find inflection points of the function f we must find points where f'' changes sign. However, because f'' is the derivative of f', any point where f'' changes sign will be a local maximum or minimum on the graph of f'.

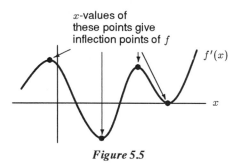

Figure 5.5

5.

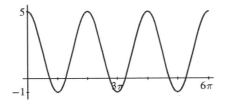

$f(x) = 2 + 3\cos x$, so $f'(x) = -3\sin x$, which is zero when $x = n\pi$ for $n - 0, 1, \ldots 5$.

The local maxima are where n is even, at $x = 0$, $x = 2\pi$, $x = 4\pi$, $x = 6\pi$. The local minima are where n is odd, at $x = \pi$, $x = 3\pi$, $x = 5\pi$. This is because when n is even, $\cos n\pi = 1$, the maximum of cosine, and so $f(x) = 5$ here, maximizing $f(x)$. When n is odd, $\cos n\pi = -1$, the minimum of cosine, and so $f(x) = -1$ here, minimizing $f(x)$..

$f''(x) = -3\cos x$, which is zero when $x = (2n+1)\pi/2$ for $n = 0, 1, \ldots 5$, so there are inflection points at $x = \pi/2, 3\pi/2, 5\pi/2, 7\pi/2, 9\pi/2$, and $11\pi/2$. These are the points with the steepest slopes, since $f'(x)$ is either maximized or minimized here.

9. We have

$$f'(x) = 10x^9 - 10$$

and

$$f''(x) = 90x^8.$$

Since $90x^8 \geq 0$ for all x, this shows that $f'(x)$ is increasing, in particular for $0 \leq x \leq 2$. Thus $f'(x)$ is maximized on the given interval when $x = 2$. Since $f'(2) = 10(2^9) - 10 = 5110 > 0$, f is increasing most rapidly when $x = 2$. Similarly, $f'(x)$ is least on the given interval when $x = 0$. $f'(0) = -10$, so this is where $f(x)$ is decreasing most rapidly.

13. The local maxima and minima of f correspond to places where f' is zero and changes sign. The points at which f changes concavity correspond to local maxima and minima of f'. The change of sign of f', from positive to negative corresponds to a maximum of f and change of sign of f' from negative to positive corresponds to a minimum of f.

17. (a) As $x \to \infty$, $f(x) \to 1$. As $x \to -\infty$, $f(x) \to 1$. As $x \to 0^+$, $f(x) \to \infty$. As $x \to 0^-$, $f(x) \to 0$.

(b) $f'(x) = \left(\frac{-1}{x^2}\right)\left(e^{\frac{1}{x}}\right)$. Thus $f'(x) < 0$ for all $x \neq 0$, which means $f(x)$ is decreasing everywhere it is defined.

(c) $f''(x) = \frac{1}{x^4}e^{\frac{1}{x}} + \frac{2}{x^3}e^{\frac{1}{x}} = \frac{(2x+1)}{x^4}e^{\frac{1}{x}}$.
$f''(x) = 0$ when $x = -1/2$.
$f''(x) < 0$ for $x < -1/2$ and $f''(x) > 0$ for $-1/2 < x < 0$ and $x > 0$.
So, $f(x)$ is concave up for $x > 0$ and $-1/2 < x < 0$, and concave down for $x < -1/2$.

(d)

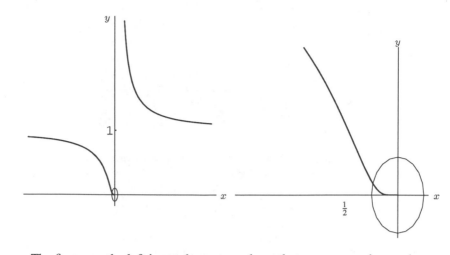

The figure on the left is not drawn to scale so that we can see the graph more globally. The figure on the right is a blow-up of the graph on the left for values of x between -1 and 0. In this picture, we have a better view of the inflection point at $x = -1/2$.

21. (a) When a number grows larger, its reciprocal grows smaller. Therefore, since f is increasing near x_0, we know that g (its reciprocal) must be decreasing. Another argument can be made using derivatives. We know that (since f is increasing) $f'(x) > 0$ near x_0. We also know (by the chain rule) that $g'(x) = (f(x)^{-1})' = -\frac{f'(x)}{f(x)^2}$. Since both $f'(x)$ and $f(x)^2$ are positive, this means $g'(x)$ is negative, which in turn means $g(x)$ is decreasing near $x = x_0$.

(b) Since f has a local maximum near x_1, $f(x)$ increases as x nears x_1, and then $f(x)$ decreases as x exceeds x_1. Thus the reciprocal of f, g, decreases as x nears x_1 and then increases as x exceeds x_1. Thus g has a local minimum at $x = x_1$. To put it another way, since f has a local maximum at $x = x_1$, we know $f'(x_1) = 0$. Since $g'(x) = -\frac{f'(x)}{f(x)^2}$, $g'(x_1) = 0$. To the left of x_1, $f'(x_1)$ is positive, so $g'(x)$ is negative. To the right of x_1, $f'(x_1)$ is negative, so $g'(x)$ is positive. Therefore, g has a local minimum at x_1.

(c) Since f is concave down at x_2, we know $f''(x_2) < 0$. We also know (from above) that

$$g''(x_2) = \frac{2f'(x_2)^2}{f(x_2)^3} - \frac{f''(x_2)f(x_2)}{f(x_2)^3} = \frac{1}{f(x_2)^2}\left(\frac{2f'(x_2)^2}{f(x_2)} - f''(x_2)\right).$$

Since $\frac{1}{f(x_2)^2} > 0$ and $2f'(x_2)^2 > 0$, and since $f(x_2) > 0$ (since f is assumed to be everywhere positive), we see that $g''(x_2)$ is positive. Thus g is concave up at x_2.

Note that for the first two parts of the problem, we didn't need to require f to be positive (only non-zero). However, it was necessary here.

25.

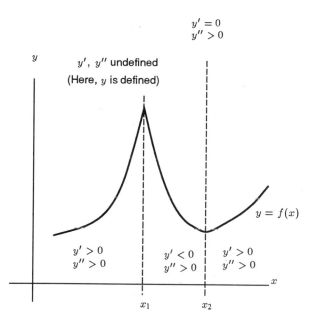

$$y' = 0$$
$$y'' > 0$$

y

y', y'' undefined
(Here, y is defined)

$y = f(x)$

$$y' > 0$$
$$y'' > 0$$

$$y' < 0$$
$$y'' > 0$$

$$y' > 0$$
$$y'' > 0$$

x

x_1 x_2

29.

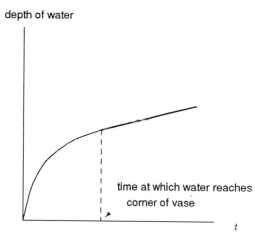

depth of water

time at which water reaches
corner of vase

t

5.3 SOLUTIONS

1. If $f'(x) > 0$ for all x then $f(x)$ is increasing everywhere. Since $f'(x) = 3x^2 + 2ax + b$, the condition is $3x^2 + 2ax + b > 0$ for all x. Since f' is positive for large $|x|$, this is the same as saying that f' has no zeros. (For f' to go from negative to positive, crossing the x-axis, would require a zero.) By the quadratic formula, that happens when the discriminant $(2a)^2 - 4(3)(b)$ is negative, that is, when $a^2 - 3b < 0$.

5. (a) $f'(x) = 4x^3 + 2ax = 2x(2x^2 + a)$; so $x = 0$ and $x = \pm\sqrt{-\frac{a}{2}}$ (if $\pm\sqrt{-\frac{a}{2}}$ is real, i.e. $-\frac{a}{2} \geq 0$) are critical points.

    (b) $x = 0$ is a critical point for any value of a. In order to guarantee that $x = 0$ is the only critical point, the factor $2x^2 + a$ should not have a root other than $x = 0$. This means $a \geq 0$, since $2x^2 + a$ has only one root ($x = 0$) for $a = 0$, and no roots for $a > 0$. There is no restriction on the constant b.

    $f''(x) = 12x^2 + 2a$. $f''(0) = 2a$.

    If $a > 0$, then by the second derivative test, $f(0)$ is a local minimum.

    If $a = 0$, then $f(x) = x^4 + b$, which has a local minimum at $x = 0$.

    So $x = 0$ is a local minimum when $a \geq 0$.

    (c) Again, b will have no effect on the location of the critical points. In order for $f'(x) = 2x(2x^2 + a)$ to have three different roots, the constant a has to be negative. Let $a = -2c^2$, for some $c > 0$. Then

    $f'(x) = 4x(x^2 - c^2) = 4x(x - c)(x + c)$.

    The critical points of f are $x = 0$ and $x = \pm c$.

    To the left of $x = -c$, $f'(x) < 0$.

    Between $x = -c$ and $x = 0$, $f'(x) > 0$.

    Between $x = 0$ and $x = c$, $f'(x) < 0$.

    To the right of $x = c$, $f'(x) > 0$.

    So, $f(-c)$ and $f(c)$ are local minima and $f(0)$ is a local maximum.

    (d) For $a \geq 0$, there is exactly one critical point, $x = 0$. For $a < 0$ there are exactly three different critical points. These exhaust all the possibilities. (Notice that the value of b is irrelevant here.)

9. (a) We have $y' = 2A(x + B)$ and $y'' = 2A$.

    (i) If A is positive, the graph concaves upward, if A is negative, the graph concaves downward.

    (ii) The larger A is in magnitude, the steeper the graph is.

    (b) B shifts the graph to the left or right depending on whether it is increased or decreased. Note that the x-intercept of the graph is at $x = -B$.

    (c) The graph is a parabola with a maximum (if A is negative) or minimum (if A is positive) at $x = -B$, where the steepness depends on the magnitude of A.

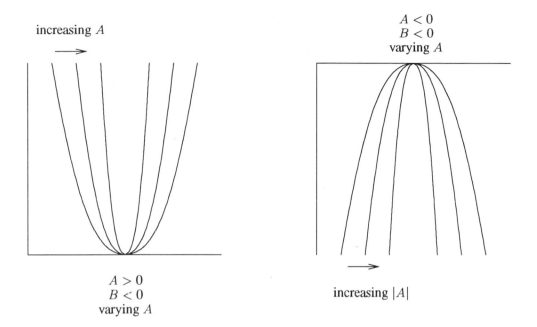

increasing A

$A < 0$
$B < 0$
varying A

$A > 0$
$B < 0$
varying A

increasing $|A|$

13. A affects the amplitude (e.g. height) of the curve. B affects the frequency. C causes a phase shift to the left or right.

Changing A:

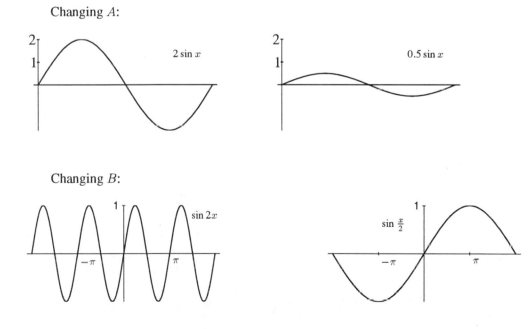

$2 \sin x$

$0.5 \sin x$

Changing B:

$\sin 2x$

$\sin \frac{x}{2}$

Changing C:

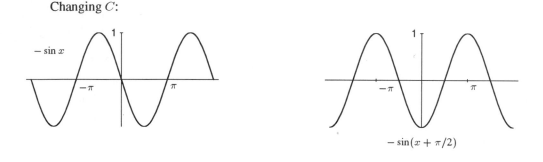

$- \sin x$

$- \sin(x + \pi/2)$

5.4 SOLUTIONS

1. (a) $N = 100 + 20x$, graphed in Figure 5.6.

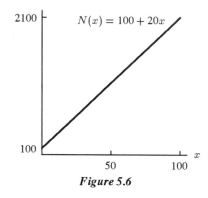

Figure 5.6

(b) $N'(x) = 20$ and its graph is just a horizontal line. This means that rate of increase of the number of bees with acres of clover is constant – each acre of clover brings 20 more bees.

On the other hand, $\frac{N(x)}{x} = \frac{100}{x} + 20$ means that the average number of bees per acre of clover approaches 20 as more acres are put under clover. As x increases, $\frac{100}{x}$ decreases to 0, so $\frac{N(x)}{x}$ approaches 20 (i.e. $\frac{N(x)}{x} \to 20$). Since the total number of bees is 20 per acre plus the original 100, the average number of bees per acre is 20 plus the 100 shared out over x acres. As x increases, the 100 are shared out over more acres, and so its contribution to the average becomes less. Thus the average number of bees per acre approaches 20 for large x.

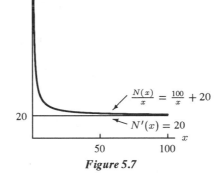

Figure 5.7

5. (a) The fixed cost is 0 because $C(0) = 0$.
    (b) Profit, $\pi(q)$, is equal to money from sales, $7q$, minus total cost to produce those items, $C(q)$.

$$\pi = 7q - 0.01q^3 + 0.6q^2 - 13q$$
$$\pi' = -0.03q^2 + 1.2q - 6 = 0$$
$$q = \frac{-1.2 \pm \sqrt{(1.2)^2 - 4(0.03)(6)}}{-0.06} \approx 5.9 \text{ min, } 34.1 \text{ max}$$

Since

$$\pi(35) = 7(35) - 0.01(35)^3 + 0.6(35)^2 - 13(35) = 245 - 148.75 = 96.25$$

$$\pi(34) = 7(34) - 0.01(34)^3 + 0.6(34)^2 - 13(34) = 238 - 141.44 = 96.56$$

The maximum profit is $\pi(34) = 96.56$. The money from sales is $238, the cost to produce the items is $141.44, resulting in a profit of $96.56.

    (c) The money from sales is equal to price×quantity sold. If the price is raised from $7 by $x to $(7 + x), the result is a reduction in sales from 34 items to $(34 - 2x)$ items. So the result of raising the price by $x is to change the money from sales from 7(34) to $(7 + x)(34 - 2x)$. If the production level is fixed at 34, then the production costs are fixed at $141.44, as found in part (b), and the profit is given by:

$$\pi(x) = (7 + x)(34 - 2x) - 141.44$$

This expression gives us the profit as a function of the change in price x, rather than as a function of quantity as found in part (b). We take the derivative of π with respect to x to find the change in price that maximizes the profit.

$$\pi' = (1)(34 - 2x) + (7 + x)(-2) = 34 - 14 - 4x = 0$$

So $x = 5$, and this must give a maximum for $\pi(x)$ since the graph of π is a parabola which opens downwards. The profit when the price is $12 (= 7 + x = 7 + 5) is thus $\pi(5) = (7 + 5)(34 - 2(5)) - 141.44 = \$146.56$. This is indeed higher than the profit when the price is $7, so the smart thing to do is to raise the price by $5.

9. It is interesting to note that to draw a graph of $C'(q)$ for this problem, you never have to know what $C(q)$ looks like, although you *could* draw a graph of $C(q)$ if you wanted to. By the formula given in the problem, we know that $C(q) = q \cdot a(q)$. Using the product rule we get that $C'(q) = a(q) + q \cdot a'(q)$.

We are given a graph of $a(q)$ which is linear, so $a(q) = b + mq$, where $b = a(0)$ is the y-intercept and m is the slope. Therefore

$$C'(q) = a(q) + q \cdot a'(q) = b + mq + q \cdot m$$
$$= b + 2mq.$$

In other words, $C'(q)$ is also linear, and it has twice the slope and the same y–intercept as $a(q)$.

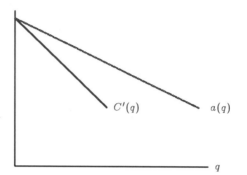

5.5 SOLUTIONS

1. We have that $v(r) = a(R - r)r^2 = aRr^2 - ar^3$, and $v'(r) = 2aRr - 3ar^2 = 2ar(R - \frac{3}{2}r)$, which is zero if $r = \frac{2}{3}R$, or if $r = 0$, and so $v(r)$ has critical points here.

$v''(r) = 2aR - 6ar$, and thus $v''(0) = 2aR > 0$, which by the second derivative test implies that v has a minimum at $r = 0$. $v''(\frac{2}{3}R) = 2aR - 4aR = -2aR < 0$, and so by the second derivative test v has a maximum at $r = \frac{2}{3}R$.

5. $f'(x) = 2(x - a_1) + 2(x - a_2) + 2(x - a_3)$.

$f'(x) = 0$ when $3x - a_1 - a_2 - a_3 = 0$, i.e. $x = \frac{a_1 + a_2 + a_3}{3}$, the average of a_1, a_2 and a_3.

This is where $f(x)$ is a minimum, since f is a parabola which opens upwards.

This is a reasonable answer since it keeps each squared term in $f(x)$ somewhat small.

9. This question implies that the line from the origin to the point $(x, R(x))$ has some relationship to $r(x)$. The slope of this line is $\frac{R(x)}{x}$, which is $r(x)$. So the point x_0 at which $r(x)$ is maximal will also be the point at which the slope of this line is maximal. The question claims that the line from the origin to $(x_0, R(x_0))$ will be tangent to the graph of $R(x)$. We can understand this by trying to see what would happen if it were otherwise.

If the line from the origin to $(x_0, R(x_0))$ intersects the graph of $R(x)$, then there are points of this graph on both sides of the line — and, in particular, there is some point x_1 such that the line from the origin to $(x_1, R(x_1))$ has larger slope than the line to $(x_0, R(x_0))$. (See the graph below.) But we picked x_0 so that no other line had larger slope, and therefore no such x_1 exists. So the original supposition is false, and the line from the origin to $(x_0, R(x_0))$ is tangent to the graph of $R(x)$.

(a)　See (b).
(b)

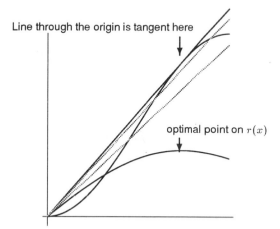

Line through the origin is tangent here

optimal point on $r(x)$

(c)

$$r(x) = \frac{R(x)}{x}$$

$$r'(x) = \frac{xR'(x) - R(x)}{x^2}$$

So when $r(x)$ is maximized $0 = \frac{R'(x)}{x} - \frac{R(x)}{x^2}$, or $R'(x) = \frac{R(x)}{x} = r(x)$, i.e. when $r(x)$ is maximized, $r(x) = R'(x)$.

Let us call the x value at which the maximum of r occurs x_m. Then the line passing through $R(x_m)$ and the origin is $y = x \cdot \frac{R(x_m)}{x_m}$. Its slope is $\frac{R(x_m)}{x_m}$, which also happens to be $r(x_m)$. In the previous paragraph, we showed that at x_m, this is also equal to the slope of the tangent to $R(x)$. So, the line through the origin *IS* the tangent line.

13. Let $y = \ln(1 + x^2)$. Then $y' = \frac{2x}{1+x^2}$. Since the denominator is always positive, the sign of y' is determined by the numerator $2x$. Thus $y' > 0$ when $x > 0$, and $y' < 0$ when $x < 0$, and we have a local (and global) minimum for y at $x = 0$. Since $y(-1) = \ln 2$ and $y(2) = \ln 5$, the global maximum is at $x = 2$. Thus $0 \le y \le \ln 5$, or (in decimals) $0 \le y < 1.61$. (Note that our upper bound has been rounded *up* from 1.6094.)

17. (a)　If, following the hint, we set $f(x) = \frac{a+x}{2} - \sqrt{ax}$, then $f(x)$ represents the difference between the arithmetic and geometric means for some fixed a and any x. We can find where this difference is minimized by solving $f'(x) = 0$. Since $f'(x) = \frac{1}{2} - \frac{1}{2}\sqrt{a}x^{-\frac{1}{2}}$, if $f'(x) = 0$ then $\frac{1}{2}\sqrt{a}x^{-\frac{1}{2}} = \frac{1}{2}$,

or $x = a$. Since $f''(x) = \frac{1}{4}\sqrt{a}x^{-\frac{3}{2}}$ is positive for all positive x, by the second derivative test $f(x)$ has a minimum at $x = a$ of $f(a) = 0$. Thus $f(x) = \frac{a+x}{2} - \sqrt{ax} \geq 0$ for all $x > 0$, which means $\frac{a+x}{2} \geq \sqrt{ax}$. Taking $x = b$, we have $\frac{a+b}{2} \geq \sqrt{ab}$. This means that the arithmetic average is greater than the geometric average unless $a = b$, in which case the two averages are equal.

Alternatively (and without using calculus): Since

$$\frac{a+b}{2} - \sqrt{ab} = \frac{a - 2\sqrt{ab} + b}{2}$$

$$= \frac{(\sqrt{a} - \sqrt{b})^2}{2} > 0,$$

$\frac{a+b}{2} > \sqrt{ab}$.

(b)   Following the hint, set $f(x) = \frac{a+b+x}{3} - \sqrt[3]{abx}$. Then $f(x)$ represents the difference between the arithmetic and geometric means for some fixed a, b and any x. We can find where this difference is minimized by solving $f'(x) = 0$. Since $f'(x) = \frac{1}{3} - \frac{1}{3}\sqrt[3]{ab}x^{-2/3}$, $f'(x) = 0$ implies that $\frac{1}{3}\sqrt[3]{ab}x^{-2/3} = \frac{1}{3}$, or $x = \sqrt{ab}$. Since $f''(x) = \frac{2}{9}\sqrt[3]{ab}x^{-5/3}$ is positive for all positive x, by the second derivative test $f(x)$ has a minimum at $x = \sqrt{ab}$. But

$$f(\sqrt{ab}) = \frac{a+b+\sqrt{ab}}{3} - \sqrt[3]{ab\sqrt{ab}} = \frac{a+b+\sqrt{ab}}{3} - \sqrt{ab} = \frac{a+b-2\sqrt{ab}}{3}.$$

By the first part of this problem, we know that $\frac{a+b}{2} - \sqrt{ab} \geq 0$, which implies that $a+b-2\sqrt{ab} \geq 0$. Thus $f(\sqrt{ab}) = \frac{a+b-2\sqrt{ab}}{3} \geq 0$. Since f has a maximum at $x = \sqrt{ab}$, $f(x)$ is always nonnegative. Thus $f(x) = \frac{a+b+x}{3} - \sqrt[3]{abx} \geq 0$, so $\frac{a+b+c}{3} \geq \sqrt[3]{abc}$. Note that equality holds only when $a = b = c$. (This may also be done without calculus, but it's harder than (a).)

5.6   SOLUTIONS

1.   Let w and l be the width and length, respectively, of the rectangular area you wish to enclose. Then

$$w + w + l = 100 \text{ feet}$$

$$l = 100 - 2w$$

$$\text{Area} = w \cdot l = w(100 - 2w) = 100w - 2w^2$$

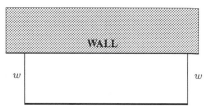

To maximize area, we solve $A' = 0$ to find critical points. This gives $A'l = 100 - 4w = 0$, so $w = 25$, $l = 50$. So the area is $25 \cdot 50 = 1250$ square feet. This is a local maximum by the second derivative test because $A'' = -4 < 0$. Since the graph of A is a parabola, the local maximum is in fact a global maximum.

5.   Let x equal the number of chairs ordered in excess of 300, so $0 \leq x \leq 100$.

$$\text{Revenue} = R = (90 - 0.25x)(300 + x)$$

$$= 27,000 - 75x + 90x - 0.25x^2 = 27,000 + 15x - 0.25x^2$$

At a critical point $dR/dx = 0$. $dR/dx = 15 - 0.5x$ so $x = 30$ gives a maximum revenue of $\$27,225$ since the graph of R is a parabola which opens downwards. The minimum is $\$0$ (when no chairs are sold).

9.   The distance from a given point on the parabola (x, x^2) to $(1, 0)$ is given by

$$D = \sqrt{(x - 1)^2 + (x^2 - 0)^2}.$$

Minimizing this is equivalent to minimizing $d = (x - 1)^2 + x^4$. (We can ignore the square root if we are only interested in minimizing because the square root is smallest when the thing it is the square root of is smallest.) To minimize d, we find its critical points by solving $d' = 0$. Since $d = (x - 1)^2 + x^4 = x^2 - 2x + 1 + x^4$,

$$d' = 2x - 2 + 4x^3$$
$$d' = 2(2x^3 + x - 1)$$

By graphing $d' = 2x^3 + x - 1$ on a calculator, we see that it has only 1 root, $x \approx 0.59$. This must give a minimum (as we can make d as large as we please); but this is confirmed by the second derivative test, as $d'' = 6x^2 + 1$, which is always positive. Thus the point $(0.59, 0.59^2) \approx (0.59, 0.35)$ is approximately the closest point of $y = x^2$ to $(1, 0)$.

13. (a) Let's suppose Bueya finds an apartment x miles from Washington University. Then her total traveling distance for the day is

$$2x + 2(14 - x) + 2|x - 5| = 2x + 28 - 2x + 2|x - 5|$$
$$= 28 + 2|x - 5|$$

Note that we had to take $|x - 5|$ as the distance to and from the bar, since we do not know whether $x \geq 5$ or $x < 5$. Since $|x - 5| \geq 0$, $28 + |x - 5|$ is minimized when $|x - 5| = 0$, i.e. when $x = 5$. So Bueya should look for an apartment as close to the bar as possible. (Is there a moral to this story?)

   (b) Now let's suppose Marie-Josée finds an apartment x miles from Washington University. Her total travel distance is

$$2x + 2(14 - x) + 2|x - 5| + 2|x - 13| = 28 + 2|x - 5| + 2|x - 13|.$$

This last sum is minimized for any x between 5 and 13. Indeed, if $5 \leq x \leq 13$, then the sum

$$28 + 2(x - 5) + 2(13 - x) = 44.$$

Thus Marie-Josée should try to live anywhere between the bar and the Gateway Arch.

5.7 SOLUTIONS

1.   (a) $f'(x) = 3x^2 + 6x + 3 = 3(x+1)^2$. Thus $f'(x) > 0$ everywhere except at $x = -1$, so it is
      increasing everywhere except perhaps at $x = -1$. The function is in fact increasing at $x = -1$
      since $f(x) > f(-1)$ for $x > -1$, and $f(x) < f(-1)$ for $x < -1$.
     (b)   The original equation can have at most one root, since it can only pass through the x-axis once if
      it never decreases. It must have one root, since $f(0) = -6$ and $f(1) = 1$.
     (c)   The root is in the interval $[0, 1]$, since $f(0) < 0 < f(1)$.
     (d)   Let $x_0 = 1$.

$$x_0 = 1$$
$$x_1 = 1 - \frac{f(1)}{f'(1)}$$
$$= 1 - \frac{1}{12} = \frac{11}{12} \approx 0.917$$
$$x_2 = \frac{11}{12} - \frac{f\left(\frac{11}{12}\right)}{f'\left(\frac{11}{12}\right)} \approx 0.913$$
$$x_3 = 0.913 - \frac{f(0.913)}{f'(0.913)} \approx 0.913.$$

Since the digits repeat, they should be accurate. Thus $x \approx 0.913$.

5.   Let $f(x) = \sin x - 1 + x$; we want to find all zeros of f, because $f(x) = 0$ implies $\sin x = 1 - x$.
     Graphing $\sin x$ and $1 - x$, we see that $f(x)$ has one solution at $x \approx \frac{1}{2}$.

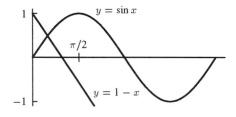

Letting $x_0 = 0.5$, and using Newton's method, we have $f'(x) = \cos x + 1$, so that

$$x_1 = 0.5 - \frac{\sin(0.5) - 1 + 0.5}{\cos(0.5) + 1} \approx 0.511,$$

$$x_2 = 0.511 - \frac{\sin(0.511) - 1 + 0.511}{\cos(0.511) + 1} \approx 0.511.$$

Thus $\sin x = 1 - x$ has one solution at $x \approx 0.511$.

9. Let $f(x) = \ln x - \frac{1}{x}$, so $f'(x) = \frac{1}{x} + \frac{1}{x^2}$.
   Now use Newton's method with an initial guess of $x_0 = 2$.

   $$x_1 = 2 - \frac{\ln 2 - \frac{1}{2}}{\frac{1}{2} + \frac{1}{4}} \approx 1.7425,$$
   $$x_2 \approx 1.763,$$
   $$x_3 \approx 1.763.$$

   Thus $x \approx 1.763$ is a solution. Since $f'(x) > 0$ for positive x, f is increasing: it must be the only solution.

13. (a) Set $f(x) = \sin x$, so $f'(x) = \cos x$. Guess $x_0 = 3$. Then

   $$x_1 = 3 - \frac{\sin 3}{\cos 3} \approx 3.1425$$
   $$x_2 \approx 3.141592653, \text{ which is correct to one billionth!}$$

   (b) Newton's method uses the tangent line at $x = 3$, i.e. $y - \sin 3 = \cos(3)(x - 3)$. Around $x = 3$, however, $\sin x$ is almost linear, since the second derivative $\sin''(\pi) = 0$. Thus using the tangent line to get an approximate value for the root gives us a very good approximation.

   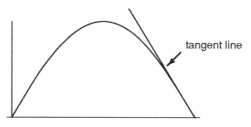

   (c) $f(x) = \sin x$

   $$[3,4]: f(3) = 0.14112$$
   $$f(4) = -0.7568$$

   The root is in

   $$[3, 3.5]: f(3.5) = -0.35078 \text{ (bisection 1)}$$
   $$[3, 3.25]: f(3.25) = -0.10819 \text{ (bisection 2)}$$
   $$[3.125, 3.25]: f(3.125) = 0.01659 \text{ (bisection 3)}$$
   $$[3.125, 3.1875]: f(3.1875) = -0.04584 \text{ (bisection 4)}$$

   We continue this process; after 11 bisections, we know the root lies between 3.1411 and 3.1416, which still is not as good an approximation as what we get from Newton's method.

SOLUTIONS TO REVIEW PROBLEMS FOR CHAPTER FIVE

1. (a) We wish to investigate the behavior of $f(x) = x^3 - 3x^2$ on the interval $-1 \le x \le 3$. We find:

$$f'(x) = 3x^2 - 6x = 3x(x - 2)$$
$$f''(x) = 6x - 6 = 6(x - 1)$$

   (b) The critical points of f are $x = 2, 0$, since $f'(x) = 0$ here. Using the second derivative test, we find that $x = 0$ is a local maximum since $f'(0) = 0$ and $f''(0) = -6 < 0$, that $x = 2$ is a local minimum since $f'(2) = 0$ and $f''(2) = 6 > 0$.
   (c) There is an inflection point at $x = 1$ since f'' changes sign at $x = 1$.
   (d) At the critical points, $f(0) = 0$ and $f(2) = -4$.
        At the endpoints: $f(-1) = -4, f(3) = 0$.
        So the global maxima are $f(0) = 0$ and $f(3) = 0$, while the global minima are $f(-1) = -4$ and $f(2) = -4$.
   (e)

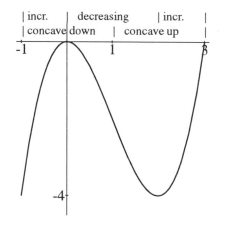

5. The polynomial $f(x)$ behaves like $2x^3$ as x goes to ∞. Therefore, $\lim\limits_{x \to \infty} f(x) = \infty$ and $\lim\limits_{x \to -\infty} f(x) = -\infty$.

   We have $f'(x) = 6x^2 - 18x + 12 = 6(x - 2)(x - 1)$, which is zero when $x = 1$ or $x = 2$.

   Also, $f''(x) = 12x - 18 = 6(2x - 3)$, which is zero when $x = \frac{3}{2}$. For $x < \frac{3}{2}$, $f''(x) < 0$; for $x > \frac{3}{2}$, $f''(x) > 0$. Thus $x = \frac{3}{2}$ is an inflection point.

   The critical points are $x = 1$ and $x = 2$, and $f(1) = 6$, $f(2) = 5$. By the second derivative test, $f''(1) = -6 < 0$, so $x = 1$ is a local maximum; $f''(2) = 6 > 0$, so $x = 2$ is a local minimum.

   Now we can draw the diagrams below.

$y' > 0$		$y' < 0$		$y' > 0$
increasing	1	decreasing	2	increasing

$y'' < 0$		$y'' > 0$
concave down	3/2	concave up

The graph of $f(x) = 2x^3 - 9x^2 + 12x + 1$ looks like this.

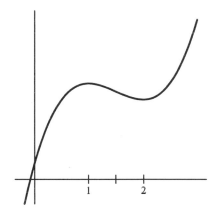

$f(x)$ has no global maximum or minimum.

9. $\lim_{x \to \infty} f(x) = +\infty$, and $\lim_{x \to -\infty} f(x) = -\infty$.
There are no asymptotes.
$f'(x) = 3x^2 + 6x - 9 = 3(x + 3)(x - 1)$. Critical points are $x = -3$, $x = 1$.
$f''(x) = 6(x + 1)$.

TABLE 5.1

x		-3		-1		1	
f'	+	0	−		−	0	+
f''	−		−	0	+		+
f	↗⌢		↘⌢		↘⌣		↗⌣

Thus, $x = -1$ is an inflection point. $f(-3) = 12$ is a local maximum; $f(1) = -20$ is a local minimum. There are no global maxima or minima.

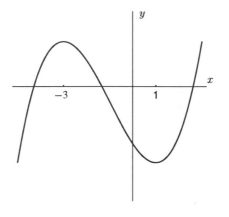

13. Since $\lim\limits_{x \to -\infty} f(x) = \lim\limits_{x \to +\infty} f(x) = 0$, $y = 0$ is a horizontal asymptote.

$f'(x) = -2xe^{-x^2}$. So, $x = 0$ is the only critical point.

$f''(x) = -2(e^{-x^2} + x(-2x)e^{-x^2}) = 2e^{-x^2}(2x^2 - 1) = 2e^{-x^2}(\sqrt{2}x - 1)(\sqrt{2}x + 1)$.

Thus, $x = \pm\frac{1}{\sqrt{2}}$ are inflection points.

TABLE 5.2

x		$\frac{-1}{\sqrt{2}}$		0		$\frac{1}{\sqrt{2}}$	
f'	$+$		$+$	0	$-$		$-$
f''	$+$	0	$-$		$-$	0	$+$
f	⌐⌣		⌐⌢		⌐⌢		⌐⌣

Thus, $f(0) = 1$ is a local and global maximum.

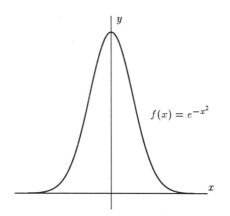

$f(x) = e^{-x^2}$

17. (a) $(-\infty, 0)$ decreasing, $(0, \infty)$ increasing.

(b) $f(0)$ is a local and global minimum.

21. (a) The concavity changes at t_1 and t_3.

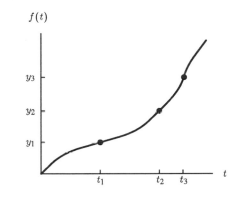

(b) $f(t)$ grows fastest where the vase is skinniest (at y_3) and slowest where the vase is widest (at y_1). The diameter of the widest part of the vase looks to be about 4 times as large as the diameter at the skinniest part. Since the area of a cross section is given by πr^2, where r is the radius, the ratio between areas of cross sections at these two places is about 4^2, so the growth rates are in a ratio of about 1 to 16 (the wide part being 16 times slower).

25. (a) $a(q)$ is represented by the slope of the line from the origin to the graph. For example, the slope of line (1) through (0,0) and $(p, C(p))$ is $\frac{C(p)}{p} = a(p)$.

(b) $a(q)$ is minimal where $a(q)$ is tangent to the graph (line (2)).

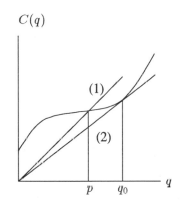

(c)   We have $a(q) = \frac{C(q)}{q}$; by the quotient rule

$$a'(q) = \frac{qC'(q) - C(q)}{q^2}$$

$$= \frac{C'(q) - \frac{C(q)}{q}}{q}$$

$$= \frac{1}{q}(C'(q) - a(q)).$$

Thus if $q = q_0$, then $a'(q_0) = \frac{1}{q}(C'(q_0) - a(q_0)) = 0$, so that $C'(q_0) = a(q_0)$; or, the average cost is minimized when it equals the marginal cost.

(d)

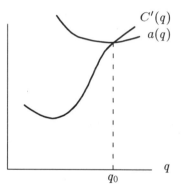

CHAPTER SIX

6.1 SOLUTIONS

1. Left-hand sum gives: $1^2(1/4) + (1.25)^2(1/4) + (1.5)^2(1/4) + (1.75)^2(1/4) = 1.96875$.
   Right-hand sum gives: $(1.25)^2(1/4) + (1.5)^2(1/4) + (1.75)^2(1/4) + (2)^2(1/4) = 2.71875$.
   We estimate the value of the integral by taking the average of these two sums, which is 2.34375.
   Since x^2 is monotonic on $1 \leq x \leq 2$, the true value of the integral lies between 1.96875 and 2.71875. Thus the most our estimate could be off is 0.375. We expect it to be much closer. (And it is — the true value of the integral is $7/3 \approx 2.333$.)

5. Since $\frac{d}{dx}(x^3 + x) = 3x^2 + 1$, by the Fundamental Theorem of Calculus, $\int_0^2 (3x^2 + 1)\, dx = (x^3 + x)\Big|_0^2 = 10$.

9. (a)

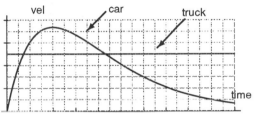

   (b) The graphs intersect twice, at about 0.7 hours and 4.3 hours. At each intersection point, the velocity of the car is equal to the velocity of the truck. If the distance apart is written as $d_{car} - d_{truck}$, then its derivative is (distance apart)$' = (d_{car})' - (d_{truck})' = v_{car} - v_{truck}$. At our intersection points $v_{car} = v_{truck}$, so (distance apart)$' = 0$. Thus these points are where the distance between the two vehicles is at a local extremum. To see this, note that from the time they start until 0.7 hours later the truck is traveling at a greater velocity than the car, so the truck is ahead of the car and pulling farther away. At 0.7 hours they are traveling at the same velocity, and after 0.7 hours the car is traveling faster than the truck, so that the car begins to gain on the truck. Thus, at 0.7 hours the truck is farther from the car than it is immediately before or after 0.7 hours. Note that this says nothing about what the distance between the two is doing outside of a small interval around 0.7 hours (later, perhaps, the truck could be even farther from the car): this is only a local extremum. Similarly, because the car's velocity is greater than the truck's after 0.7 hours, it will catch up with the truck and eventually pass and pull away from the truck until 4.3 hours, at which point the two are again traveling at the same velocity. After 4.3 hours the truck travels faster than the car, so that it now gains on the car. Thus, 4.3 hours represents the point where the car is farthest ahead of the truck.

13. (a) For the first twelve months, the total number of appliances sold is

$$7 + 9 + 11 + \cdots + 29 = 216,$$

so the average number $= \frac{216}{12} = 18$ appliances per month.

(b)   Average $= \dfrac{1}{12} \displaystyle\int_0^{12} (2t + 5)\, dt$, which by the Fundamental Theorem is given by $\dfrac{1}{12}(t^2 + 5t)\Big|_0^{12} =$

$\dfrac{204}{12} = 17$ appliances per month.

(c)   They are close, but not equal. Using integration gives an underestimate of the true value. This is because the function $2t + 5$ equals the rate at which she sells only at the end of each month; see the figure below.

(d)   The integral is easier to calculate than the sum, particularly for a large number of months.

(e)   The rectangles represent the true answer. The shaded region is the error.

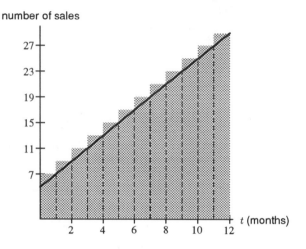

6.2   SOLUTIONS

1. $\int_1^3 (x^2 - x)\, dx = \int_1^3 x^2\, dx - \int_1^3 x\, dx.$

$\int_1^3 3x^2\, dx = 26$, so $\int_1^3 x^2\, dx = 26/3$, since $\int_1^3 3x^2 dx = 3\int_1^3 x^2 dx.$

$\int_1^3 2x\, dx = 8$, so $\int_1^3 x\, dx = 4$, since $\int_1^3 2x dx = 2\int_1^3 x dx.$

Thus, $\int_1^3 (x^2 - x)\, dx = \frac{26}{3} - 4 = \frac{14}{3}.$

5.   (a) $\dfrac{1}{\sqrt{2\pi}} \displaystyle\int_1^3 e^{-\frac{x^2}{2}}\, dx$

$= \dfrac{1}{\sqrt{2\pi}} \displaystyle\int_0^3 e^{-\frac{x^2}{2}}\, dx - \dfrac{1}{\sqrt{2\pi}} \displaystyle\int_0^1 e^{-\frac{x^2}{2}}\, dx$

$\approx 0.4987 - 0.3413 = 0.1574.$

(b) $\left(\text{by symmetry of } e^{x^2/2}\right)$ $\dfrac{1}{\sqrt{2\pi}} \displaystyle\int_{-2}^3 e^{-\frac{x^2}{2}}\, dx = \dfrac{1}{\sqrt{2\pi}} \displaystyle\int_{-2}^0 e^{-\frac{x^2}{2}}\, dx + \dfrac{1}{\sqrt{2\pi}} \displaystyle\int_0^3 e^{-\frac{x^2}{2}}\, dx$

$= \dfrac{1}{\sqrt{2\pi}} \displaystyle\int_0^2 e^{-\frac{x^2}{2}}\, dx + \dfrac{1}{\sqrt{2\pi}} \displaystyle\int_0^3 e^{-\frac{x^2}{2}}\, dx$

$\approx 0.4772 + 0.4987 = 0.9759.$

9.   (a)

TABLE 6.1

x	0	1	2	3	4	5
$F(x)$	0	1	4	9	16	25

(b) F is clearly increasing for $x > 0$; as x increases, so does the area under the curve. F is also concave up for $x > 0$ — as x gets bigger the size of the area added to the integral gets bigger as well.

(c) $F(-1) = \int_0^{-1} 2t\, dt = -(\int_{-1}^0 2t\, dt)$. The quantity inside the parentheses represents the area between $y = 2t$ and the t-axis for $-1 \le t \le 0$. This area is below the t-axis, hence it is negative. So $F(-1) = -(\text{negative area})$. Thus $F(-1)$ is positive.

6.3 **SOLUTIONS**

1.

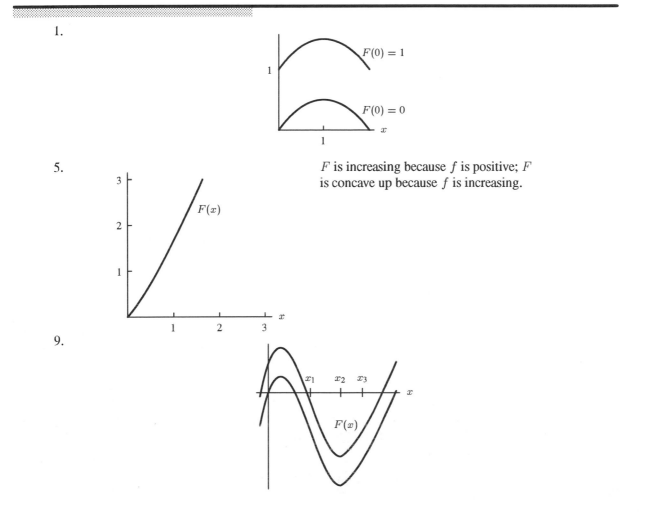

5. F is increasing because f is positive; F is concave up because f is increasing.

9.

Note that since $f(x_2) = 0$, $f'(x_2) > 0$, so $F(x_2)$ is a local minimum. Since $f'(x_1) = 0$ and f changes from decreasing to increasing at $x = x_1$, F has an inflection point at $x = x_1$.

13. Let $y'(t) = \frac{dy}{dt}$. Then y is the antiderivative of y' such that $y(0) = 0$. We know that

$$y(x) = \int_0^x y'(t)\, dt.$$

Thus, $y(x)$ is the area under the graph of $\frac{dy}{dt}$ from $t = 0$ to $t = x$ (note: we interpret "area" to be negative if a region lies below the t-axis). We therefore know that $y(t_1) = 2$, $y(t_3) = 2 - 2 = 0$, and $y(t_5) = 2$.

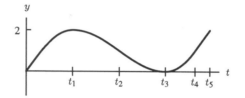

The function y' is positive on the intervals $(0, t_1)$ and (t_3, ∞), so y is increasing on those intervals. y' is negative on the interval (t_1, t_3), so y is decreasing on that interval. y' is increasing on the interval (t_2, t_4), so y is concave up on that interval; y' is decreasing on $(0, t_2)$, so y is concave down there. t_2, the point where the concavity changes, is an inflection point. Finally, since y' is constant on the interval (t_4, ∞), y's graph is linear with positive slope on this interval. $y(t_1) = 2$ is a local maximum, and $y(t_3) = 0$ is a local minimum.

17. Let's start by making sure that we know what the graph is telling us. This car starts by going 50 mph for two hours, turns around quickly and goes 50 mph in the opposite direction for one hour, turns around again, and goes 50 mph for another hour in the original direction. The distance, D, it has moved by time T is given by the integral of the velocity up to that time:

$$D(T) = \int_0^T v(t)\, dt.$$

We calculate this integral from the area under the curve, with area below the curve being subtracted. For the first two hours the distance moved is given by the area shown in Figure 6.1; this area has magnitude $50T$. Thus

$$D(T) = 50T \qquad \text{for } 0 \le T < 2.$$

At the end of the first two hours, the car is 100 miles from its starting point. If T is between 2 and 3 hours, the car has moved forwards for 2 hours, and backwards for the remainder of the time, namely $(T - 2)$ hours. (See Figure 6.2.) Thus the car has gone forward 100 miles, and backwards $50(T - 2)$ miles. The total distance from its starting point is given by

$$D(T) = 100 - 50(T - 2) \qquad \text{for } 2 \le T < 3$$

which simplifies to

$$D(T) = 200 - 50T \qquad \text{for } 2 \leq T < 3.$$

By the end of three hours, the car is $100 - 50 = 50$ miles from its starting point. During the last hour, the car travels in the original direction again. See Figure 6.3. For T over 3 hours, the car starts 50 miles from its starting point and then moves in the original direction for another $(T - 3)$ hours, covering an additional $50(T - 3)$ miles. Thus

$$D(T) = 50 + 50(T - 3) \qquad \text{for } 3 \leq T \leq 4,$$

giving

$$D(T) = 200 - 50T \qquad \text{for } 3 \leq T \leq 4.$$

A graph of the function $D(T)$ is in Figure 6.4.

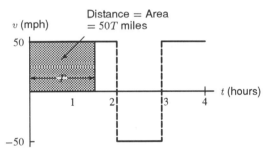

Figure 6.1: During the first 2 hours

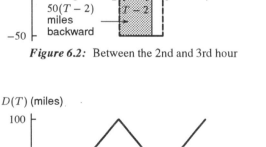

Figure 6.2: Between the 2nd and 3rd hour

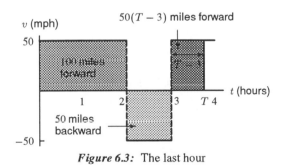

Figure 6.3: The last hour

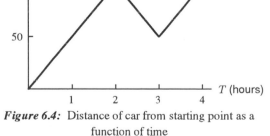

Figure 6.4: Distance of car from starting point as a function of time

6.4 SOLUTIONS

1. $5x$

5. $\sin t$

9. $-\dfrac{1}{2z^2}$

13. $\dfrac{t^4}{4} - \dfrac{t^3}{6} - \dfrac{t^2}{2}$

17. $-\cos 2\theta$

21. $\sin t + \tan t$

25. $e^2y + \dfrac{2^y}{\ln 2}$

29. The general antiderivative of $f(x)$ is $F(x) = 3x + C$. Since $F(0) = 2$, we have $F(0) = 3(0) + C = C = 2$. Thus $C = 2$, and $F(x) = 3x + 2$.

33. The general antiderivative of $f(x)$ is $F(x) = -\cos x + C$. Since $F(0) = 2$, we have $F(0) = -\cos 0 + C = -1 + C = 2$. Thus $C = 3$, and $F(x) = -\cos x + 3$.

37. $\sin \theta + C$

41. $\pi x + \dfrac{x^{12}}{12} + C$

45. $\displaystyle \int \frac{1}{e^z}\, dz = \int e^{-z}\, dz = -e^{-z} + C$

49. (a)

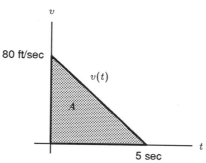

(b) The total distance is represented by the shaded region A, the area under the graph of $v(t)$.
(c) The area A, a triangle, is given by

$$A = \frac{1}{2}(\text{base})(\text{height}) = \frac{1}{2}(5\,\text{sec})(80\,\text{ft/sec}) = 200\,\text{ft}.$$

(d) Using integration and the Fundamental Theorem of Calculus, we have $A = \int_0^5 v(t)\, dt$ or $A = s(5) - s(0)$, where $s(t)$ is an antiderivative of $v(t)$.

We have that $a(t)$, the acceleration, is constant: $a(t) = k$ for some constant k. Therefore

$v(t) = kt + C$ for some constant C. We have $80 = v(0) = k(0) + C = C$, so that $v(t) = kt + 80$. Putting in $t = 5, 0 = v(5) = (k)(5) + 80$, or $k = -80/5 = -16$.

Thus $v(t) = -16t + 80$, and an antiderivative for $v(t)$ is $s(t) = -8t^2 + 80t + C$. Since the total distance traveled at $t = 0$ is 0, we have $s(0) = 0$ which means $C = 0$. Finally, $A = \int_0^5 v(t)\, dt = s(5) - s(0) = (-8(5)^2 + (80)(5)) - (-8(0)^2 + (80)(0)) = 200$ ft, which agrees with the previous part.

53. (a) $a(t) = 1.6$, so $v(t) = 1.6t + v_0 = 1.6t$, since the initial velocity is 0.

(b) $s(t) = 0.8t^2 + s_0$

6.5 SOLUTIONS

1. The velocity as a function of time is given by: $v = v_0 + at$. Since the object starts from rest, $v_0 = 0$, and the velocity is just the acceleration times time: $v = -32t$. Integrating this, we get position as a function of time: $y = -16t^2 + y_0$, where the last term, y_0, is the initial position at the top of the tower, so $y_0 = 400$ feet. Thus we have a function giving position as a function of time: $y = -16t^2 + 400$.

To find at what time the object hits the ground, we find t when $y = 0$. We solve $0 = -16t^2 + 400$ for t, getting $t^2 = \frac{400}{16} = 25$, so $t = 5$. Therefore the object hits the ground after 5 seconds. At this time it is moving with a velocity $v = -32(5) = -160$ feet/second.

5. Let the acceleration due to gravity equal $-k$ meters/sec^2, for some positive constant k, and suppose the object falls from an initial height of $s(0)$ meters.

We have $a(t) = \frac{dv}{dt} = -k$, so that $v(t) = -kt + v_0$. Since the initial velocity is zero, we have $v(0) = -k(0) + v_0 = 0$, which means $v_0 = 0$. Our formula becomes $v(t) = \frac{ds}{dt} = -kt$. This means $s(t) = \frac{-kt^2}{2} + s_0$. Since $s(0) = \frac{-k(0)^2}{2} + s_0$, we have $s_0 = s(0)$, and our formula becomes $s(t) = \frac{-kt^2}{2} + s(0)$. Suppose that the object falls for t seconds. Assuming it hasn't hit the ground, its height is $\frac{-kt^2}{2} + s(0)$, so that the distance traveled is $s(0) - (\frac{-kt^2}{2} + s(0)) = \frac{kt^2}{2}$ meters, which is proportional to t^2.

SOLUTIONS TO REVIEW PROBLEMS FOR CHAPTER SIX

1. True. The antiderivatives of $3x^2$ are of the form $x^3 + C$. No two such curves intersect. (If they did, then we'd have $x^3 + C = x^3 + C'$, so $C = C'$, but then the curves are the same!)

5. $f(x) = \frac{1}{2}x^4 - \ln |x| + \frac{1}{x} + C$.

9.

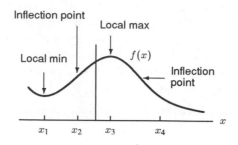

13. Antiderivative $G(t) = 5t + \sin t$

17. Antiderivative $H(t) = t - 2\ln|t| - 1/t$.

21. $e^x + e^{1+x} + C$

25. $\dfrac{(1 + \sin t)^{30}}{30} + C$

29. $3e^x + \dfrac{2^x}{\ln 2} + C$

33. $\displaystyle\int \sqrt{x}\left(1 - \frac{1}{x^{\frac{3}{2}}}\right)\,dx = \int (x^{1/2} - \frac{1}{x})\,dx = \frac{2}{3}x^{3/2} - \ln|x| + C$

37. (a) Using $g = -32$ ft/sec^2, we have

TABLE 6.2

t (sec)	0	1	2	3	4	5
$v(t)$ (ft/sec)	80	48	16	−16	−48	−80

(b) The object reaches its highest point when $v = 0$, which appears to be at $t = 2.5$ seconds. By symmetry, the object should hit the ground again at $t = 5$ seconds.

(c) Left sum: $80(1) + 48(1) + 16(\frac{1}{2}) = 136$ ft.

Right sum: $48(1) + 16(1) + (-16)\frac{1}{2} = 56$ ft.

The left sum is an overestimate, the right sum an underestimate.

(d) We have $v(t) = 80 - 32t$, so antidifferentiation yields $s(t) = 80t - 16t^2 + s_0$.

But $s_0 = 0$, so $s(t) = 80t - 16t^2$.

At $t = 2.5$, $s(t) = 100$ ft., so 100 ft. is the highest point.

41. (a) In the beginning, both birth and death rates are small; this is consistent with a very small population. Both rates begin climbing, the birth rate faster than the death rate, which is consistent with a growing population. The death rate is then high, but it begins to decrease as the population decreases.

(b)

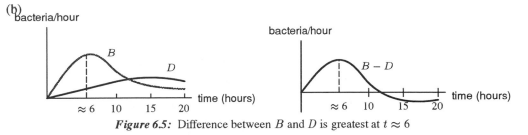

Figure 6.5: Difference between B and D is greatest at $t \approx 6$

The bacteria population is growing most quickly when $B-D$, the rate of change of population, is maximal; that happens when B is farthest above D, which is at a point where the slopes of both graphs are equal. The point on this graph satisfying that criterion is $t \approx 6$, so the greatest rate of increase occurs about 6 hours after things have begun.

(c)   Total number born by time t is the area under the B graph from $t = 0$ up to time t.

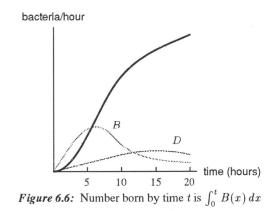

Figure 6.6: Number born by time t is $\int_0^t B(x)\, dx$

Total number alive at time t is the number born minus the number died, which is the area under the B graph up to the time t, minus the area under the D graph up to time t.

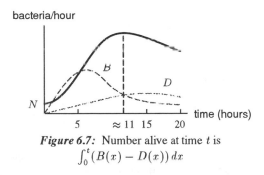

Figure 6.7: Number alive at time t is
$$\int_0^t (B(x) - D(x))\, dx$$

From Figure 6.7, we see that the population is at a maximum when $B = D$, that is, after about 11 hours. This stands to reason, because $B - D$ is the rate of change of population, so population is maximized when $B - D = 0$, that is, when $B = D$.

44. (a) $a(T) = \frac{1}{T} \int_0^T \sin t \, dt = \frac{1}{T}(1 - \cos T)$.
     (b)

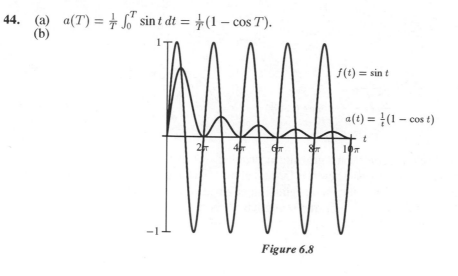

Figure 6.8

$$a'(T) = \frac{d}{dT}\left(\frac{1}{T}\int_0^T f(t)\,dt\right)$$

$$= \frac{d}{dT}\left(\frac{1}{T}\right)\int_0^T f(t)\,dt + \frac{1}{T}\frac{d}{dT}\left(\int_0^T f(t)\,dt\right).$$

Using the Fundamental Theorem of Calculus,

$$= -\frac{1}{T^2}\int_0^T f(t)\,dt + \frac{1}{T}f(T)$$

$$= -\frac{a(T)}{T} + \frac{f(T)}{T}.$$

So $a'(T) > 0$ (a is increasing) if $a(T) < f(T)$, and $a'(T) < 0$ (a is decreasing) if $a(T) > f(T)$. Thus $a(T)$ is maximum or minimum when $a(T) = f(t)$.

45. (a) We integrate by parts. Let $u = t$, so $u' = 1$, and let $v' = \sin t$, so that $v = -\cos t$. We obtain

$$a(T) = \frac{1}{T}\int_0^T t\sin t\,dt = \frac{1}{T}\left[(-t\cos t)\Big|_0^T + \int_0^T \cos t\,dt\right] = -\cos T + \frac{\sin T}{T}.$$

(b) As in Problem 44, we know that $a(t)$ is increasing when $a(t) < f(t)$ and decreasing when $a(t) > f(t)$. We also know that $a(t)$ is maximal or minimal when $a(t) = f(t)$.

So we really only need to know where the graphs of a and f cross. Looking at the graph tells us that they cross just a little before each multiple of π. One could find the actual points by computing where $a'(t) = 0$.

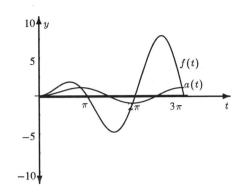

CHAPTER SEVEN

7.1 SOLUTIONS

1. $f(x) = 3$, so $F(x) = 3x + C$. $F(0) = 0$ implies that $3 \cdot 0 + C = 0$, so $C = 0$. Thus $F(x) = 3x$ is the only possibility.

5. $f(x) = x^2$, so $F(x) = \frac{x^3}{3} + C$. $F(0) = 0$ implies that $\frac{0^3}{3} + C = 0$, so $C = 0$. Thus $F(x) = \frac{x^3}{3}$ is the only possibility.

9. $f(x) = \sin x$, so $F(x) = -\cos x + C$. $F(0) = 0$ implies that $-\cos 0 + C = 0$, so $C = 1$. Thus $F(x) = -\cos x + 1$ is the only possibility.

13. One antiderivative is $7t - \frac{t^9}{72} + \ln|t|$. The general form for all antiderivatives is
$$7t - \frac{t^9}{72} + \ln|t| + C.$$

17. Let $F(x) = \frac{x^2 \sin x}{2}$. Then by the product rule $F'(x) = x \sin x + \frac{x^2 \cos x}{2} \neq x \cos x$. Thus $F(x)$ is not an antiderivative of $x \cos x$.

21. $3 \ln|t| + \dfrac{2}{t} + C$

25. Since $f(x) = x + 1 + \frac{1}{x}$, the indefinite integral is $\frac{1}{2}x^2 + x + \ln|x| + C$

29. $2e^x - 8 \sin x + C$

33. $\int (x+1)^3 \, dx = \frac{(x+1)^4}{4} + C$.
Another way to work the problem is to expand $(x+1)^3$ to $x^3 + 3x^2 + 3x + 1$:

$$\int (x+1)^3 \, dx = \int (x^3 + 3x^2 + 3x + 1) \, dx = \frac{x^4}{4} + x^3 + \frac{3}{2}x^2 + x + C.$$

It can be shown that these answers are the same by expanding $\frac{(x+1)^4}{4}$.

37. $e^{5+x} + \frac{1}{5}e^{5x} + C$, since $\frac{d}{dx}(e^{5x}) = 5e^{5x}$.

41. $\int_2^5 (x^3 - \pi x^2) \, dx = \left(\frac{x^4}{4} - \frac{\pi x^3}{3} \right) \Big|_2^5 = \frac{609}{4} - 39\pi \approx 29.728$.

45. $\int_0^{\pi/4} (\sin t + \cos t) \, dt = (-\cos t + \sin t) \Big|_0^{\pi/4} = \left(-\frac{\sqrt{2}}{2} + 1 + \frac{\sqrt{2}}{2} \right) = 1$.

49. $\int 2^x \, dx = \frac{1}{\ln 2} 2^x + C$, since $\frac{d}{dx} 2^x - \ln 2 \cdot 2^x$, so
$$\int_{-1}^1 2^x \, dx = \frac{1}{\ln 2} \left[2^x \Big|_{-1}^1 \right] = \frac{3}{2 \ln 2} \approx 2.164.$$

53.   The average value of $v(x)$ on the interval $1 \leq x \leq c]$ is

$$\frac{1}{c-1} \int_1^c \frac{6}{x^2}\, dx = \frac{1}{c-1} \left(-\frac{6}{x} \right)\Big|_1^c = \frac{1}{c-1} \left(\frac{-6}{c} + 6 \right) = \frac{6}{c}.$$

Since $\dfrac{1}{c-1} \displaystyle\int_1^c \frac{6}{x^2}\, dx = 1$, $\frac{6}{c} = 1$, so $c = 6$.

57.   (a)

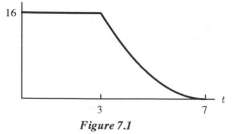

Figure 7.1

(b)   7 years, because $t^2 - 14t + 49 = (t-7)^2$ indicates that the rate of flow was zero after 7 years.

(c)

$$\text{Area under the curve} = 3(16) + \int_3^7 (t^2 - 14t + 49)\, dt$$

$$= 48 + \left(\frac{1}{3}t^3 - 7t^2 + 49t \right)\Big|_3^7$$

$$= 48 + \frac{343}{3} - 343 + 343 - 9 + 63 - 147$$

$$= \frac{208}{3} = 69\frac{1}{3}.$$

7.2 SOLUTIONS

1.   (a) $\frac{d}{dx} \sin(x^2 + 1) = 2x \cos(x^2 + 1)$; $\frac{d}{dx} \sin(x^3 + 1) = 3x^2 \cos(x^3 + 1)$
      (b)   (i) $\frac{1}{2} \sin(x^2 + 1) + C$ (ii) $\frac{1}{3} \sin(x^3 + 1) + C$
      (c)   (i) $-\frac{1}{2} \cos(x^2 + 1) + C$ (ii) $-\frac{1}{3} \cos(x^3 + 1) + C$

5.   Make the substitution $w = \sin x$, $dw = \cos x\, dx$. We have
$$\int e^{\sin x} \cos x\, dx = \int e^w\, dw = e^w + C = e^{\sin x} + C.$$

9.   For Problem 12 we use the substitution $w = -x^2$, $dw = -2x\, dx$.
      For Problem 13 we use the substitution $w = y^2 + 5$, $dw = 2y\, dy$.

For Problem 14 we use the substitution $w = t^3 - 3$, $dw = 3t^2 \, dt$.
For Problem 15 we use the substitution $w = x^2 - 4$, $dw = 2x \, dx$.
For Problem 16 we use the substitution $w = y + 5$, $dw = dy$.
For Problem 17 we use the substitution $w = 2t - 7$, $dw = 2 \, dt$.
For Problem 18 we use the substitution $w = x^2 + 3$, $dw = 2x \, dx$.
For Problem 19 it would be easier if we just multiply out $(x^2 + 3)^2$ and then integrate.
For Problem 20 use the substitution $w = 4 - x$, $dw = -dx$.
For Problem 21 use the substitution $w = \cos\theta + 5$, $dw = -\sin\theta \, d\theta$.
For Problem 22 use the substitution $w = x^3 + 1$, $dw = 3x^2 dx$.
For Problem 23 use the substitution $w = \sin\alpha$, $dw = \cos\alpha \, d\alpha$.

12. We use the substitution $w = -x^2$, $dw = -2x \, dx$.

$$\int xe^{-x^2} \, dx = -\frac{1}{2} \int e^{-x^2}(-2x \, dx) = -\frac{1}{2} \int e^w \, dw$$

$$= -\frac{1}{2}e^w + C = -\frac{1}{2}e^{-x^2} + C.$$

Check: $\frac{d}{dx}(-\frac{1}{2}e^{-x^2} + C) = (-2x)(-\frac{1}{2}e^{-x^2}) = xe^{-x^2}.$

13. We use the substitution $w = y^2 + 5$, $dw = 2y \, dy$.

$$\int y(y^2 + 5)^8 \, dy = \frac{1}{2} \int (y^2 + 5)^8 (2y \, dy)$$

$$= \frac{1}{2} \int w^8 \, dw = \frac{1}{2}\frac{w^9}{9} + C$$

$$= \frac{1}{18}(y^2 + 5)^9 + C.$$

Check: $\frac{d}{dy}(\frac{1}{18}(y^2 + 5)^9 + C) = \frac{1}{18}[9(y^2 + 5)^8(2y)] = y(y^2 + 5)^8.$

14. We use the substitution $w = t^3 - 3$, $dw = 3t^2 \, dt$.

$$\int t^2(t^3 - 3)^{10} \, dt - \frac{1}{3} \int (t^3 - 3)^{10}(3t^2 dt) = \int w^{10} \left(\frac{1}{3} \, dw\right)$$

$$= \frac{1}{3}\frac{w^{11}}{11} + C = \frac{1}{33}(t^3 - 3)^{11} + C.$$

Check: $\frac{d}{dt}[\frac{1}{33}(t^3 - 3)^{11} + C] = \frac{1}{3}(t^3 - 3)^{10}(3t^2) = t^2(t^3 - 3)^{10}.$

15. We use the substitution $w = x^2 - 4$, $dw = 2x \, dx$.

$$\int x(x^2 - 4)^{\frac{7}{2}} \, dx = \frac{1}{2} \int (x^2 - 4)^{\frac{7}{2}}(2x dx) = \frac{1}{2} \int w^{\frac{7}{2}} \, dw$$

$$= \frac{1}{2}(\frac{2}{9}w^{\frac{9}{2}}) + C = \frac{1}{9}(x^2 - 4)^{\frac{9}{2}} + C.$$

Check: $\frac{d}{dx}[\frac{1}{9}(x^2-4)^{\frac{9}{2}}+C] = \frac{1}{9}\left[\frac{9}{2}(x^2-4)^{\frac{7}{2}}\right]2x = x(x^2-4)^{\frac{7}{2}}$.

16.   We use the substitution $w = y+5$, $dw = dy$, to get

$$\int \frac{dy}{y+5} = \int \frac{dw}{w} = \ln|w| + C = \ln|y+5| + C.$$

Check: $\frac{d}{dy}(\ln|y+5|+C) = \frac{1}{y+5}$.

17.   We use the substitution $w = 2t-7$, $dw = 2\,dt$.

$$\int (2t-7)^{73}\,dt = \frac{1}{2}\int w^{73}\,dw = \frac{1}{(2)(74)}w^{74} + C = \frac{1}{148}(2t-7)^{74} + C.$$

Check: $\frac{d}{dt}\left[\frac{1}{148}(2t-7)^{74}+C\right] = \frac{74}{148}(2t-7)^{73}(2) = (2t-7)^{73}$.

18.   We use the substitution $w = x^2+3$, $dw = 2x\,dx$.

$$\int x(x^2+3)^2\,dx = \int w^2(\frac{1}{2}\,dw) = \frac{1}{2}\frac{w^3}{3} + C = \frac{1}{6}(x^2+3)^3 + C.$$

Check: $\frac{d}{dx}\left[\frac{1}{6}(x^2+3)^3+C\right] = \frac{1}{6}\left[3(x^2+3)^2(2x)\right] = x(x^2+3)^2$.

19.   In this case, it seems easier not to substitute.

$$\int (x^2+3)^2\,dx = \int (x^4+6x^2+9)\,dx = \frac{x^5}{5} + 2x^3 + 9x + C.$$

Check: $\frac{d}{dx}\left[\frac{x^5}{5}+2x^3+9x+C\right] = x^4+6x^2+9 = (x^2+3)^2$.

20.   We use the substitution $w = 4-x$, $dw = -dx$.

$$\int \frac{1}{\sqrt{4-x}}\,dx = -\int \frac{1}{\sqrt{w}}\,dw = -2\sqrt{w} + C = -2\sqrt{4-x} + C.$$

Check: $\frac{d}{dx}(-2\sqrt{4-x}+C) = -2\cdot\frac{1}{2}\cdot\frac{1}{\sqrt{4-x}}\cdot-1 = \frac{1}{\sqrt{4-x}}$.

21.   We use the substitution $w = \cos\theta + 5$, $dw = -\sin\theta\,d\theta$.

$$\int \sin\theta(\cos\theta+5)^7\,d\theta = -\int w^7\,dw = -\frac{1}{8}w^8 + C$$
$$= -\frac{1}{8}(\cos\theta+5)^8 + C.$$

Check:

$$\frac{d}{d\theta}\left[-\frac{1}{8}(\cos\theta+5)^8+C\right] = -\frac{1}{8}\cdot 8(\cos\theta+5)^7\cdot(-\sin\theta)$$
$$= \sin\theta(\cos\theta+5)^7$$

22. We use the substitution $w = x^3+1$, $dw = 3x^2\,dx$, to get

$$\int x^2 e^{x^3+1}\,dx = \frac{1}{3}\int e^w\,dw = \frac{1}{3}e^w + C = \frac{1}{3}e^{x^3+1}+C.$$

Check: $\frac{d}{dx}\left(\frac{1}{3}e^{x^3+1}+C\right) = \frac{1}{3}e^{x^3+1}\cdot 3x^2 = x^2 e^{x^3+1}.$

23. We use the substitution $w = \sin\alpha$, $dw = \cos\alpha\,d\alpha$.

$$\int \sin^3\alpha\cos\alpha\,d\alpha = \int w^3\,dw = \frac{w^4}{4}+C = \frac{\sin^4\alpha}{4}+C.$$

Check: $\frac{d}{d\alpha}\left(\frac{\sin^4\alpha}{4}+C\right) = \frac{1}{4}\cdot 4\sin^3\alpha\cdot\cos\alpha = \sin^3\alpha\cos\alpha.$

25. We use the substitution $w = \ln z$, $dw = \frac{1}{z}\,dz$.

$$\int \frac{(\ln z)^2}{z}\,dz = \int w^2\,dw = \frac{w^3}{3}+C = \frac{(\ln z)^3}{3}+C.$$

Check: $\frac{d}{dz}\left[\frac{(\ln z)^3}{3}+C\right] = 3\cdot\frac{1}{3}(\ln z)^2\cdot\frac{1}{z} = \frac{(\ln z)^2}{z}.$

29. We use the substitution $w = e^t+t$, $dw = (e^t+1)\,dt$.

$$\int \frac{e^t+1}{e^t+t}\,dt = \int \frac{1}{w}\,dw = \ln|w|+C = \ln|e^t+t|+C.$$

Check: $\frac{d}{dt}(\ln|e^t+t|+C) = \frac{e^t+1}{e^t+t}.$

33. We use the substitution $w = \sqrt{y}$, $dw = \frac{1}{2\sqrt{y}}\,dy$.

$$\int \frac{e^{\sqrt{y}}}{\sqrt{y}}\,dy = 2\int e^w\,dw = 2e^w+C = 2e^{\sqrt{y}}+C.$$

Check: $\frac{d}{dy}(2e^{\sqrt{y}}+C) = 2e^{\sqrt{y}}\cdot\frac{1}{2\sqrt{y}} = \frac{e^{\sqrt{y}}}{\sqrt{y}}.$

37.   We use the substitution $w = 1 + 2x^3$, $dw = 6x^2\, dx$.

$$\int x^2(1 + 2x^3)^2\, dx = \int w^2(\tfrac{1}{6}\, dw) = \tfrac{1}{6}(\tfrac{w^3}{3}) + C = \tfrac{1}{18}(1 + 2x^3)^3 + C.$$

Check: $\dfrac{d}{dx}\left[\tfrac{1}{18}(1 + 2x^2)^3 + C\right] = \tfrac{1}{18}[3(1 + 2x^3)^2(6x^2)] = x^2(1 + 2x^3)^2.$

41.   We use the substitution $w = e^x + e^{-x}$, $dw = (e^x - e^{-x})\, dx$.

$$\int \frac{e^x - e^{-x}}{e^x + e^{-x}}\, dx = \int \frac{dw}{w} = \ln|w| + C = \ln(e^x + e^{-x}) + C.$$

(We can drop the absolute value signs since $e^x + e^{-x} > 0$ for all x).

Check: $\dfrac{d}{dx}[\ln(e^x + e^{-x}) + C] = \dfrac{1}{e^x + e^{-x}}(e^x - e^{-x}).$

45.   Since $v = \dfrac{dh}{dt}$, it follows that $h(t) = \displaystyle\int v(t)\, dt$ and $h(0) = h_0$. Since

$$v(t) = \frac{mg}{k}\left(1 - e^{-\frac{k}{m}t}\right) = \frac{mg}{k} - \frac{mg}{k}e^{-\frac{k}{m}t},$$

we have

$$h(t) = \int v(t)\, dt = \frac{mg}{k}\int dt - \frac{mg}{k}\int e^{-\frac{k}{m}t}\, dt.$$

The first integral is simply $\dfrac{mg}{k}t + C$. To evaluate the second integral, make the substitution $w = -\dfrac{k}{m}t$. Then

$$dw = -\frac{k}{m}\, dt,$$

so

$$\int e^{-\frac{k}{m}t}\, dt = \int e^w\left(-\frac{m}{k}\right)\, dw = -\frac{m}{k}e^w + C = -\frac{m}{k}e^{-\frac{k}{m}t} + C.$$

Thus

$$h(t) = \int v\, dt = \frac{mg}{k}t - \frac{mg}{k}\left(-\frac{m}{k}e^{-\frac{k}{m}t}\right) + C$$

$$= \frac{mg}{k}t + \frac{m^2 g}{k^2}e^{-\frac{k}{m}t} + C.$$

Since $h(0) = h_0$,

$$h_0 = \frac{mg}{k}\cdot 0 + \frac{m^2 g}{k^2}e^0 + C;$$

$$C = h_0 - \frac{m^2 g}{k^2}.$$

Thus

$$h(t) = \frac{mg}{k}t + \frac{m^2 g}{k^2}e^{-\frac{k}{m}t} - \frac{m^2 g}{k^2} + h_0$$

$$h(t) = \frac{mg}{k}t - \frac{m^2 g}{k^2}\left(1 - e^{-\frac{k}{m}t}\right) + h_0.$$

7.3 SOLUTIONS

1. (a) We substitute $w = 1 + x^2$, $dw = 2x\,dx$.

$$\int_{x=0}^{x=1} \frac{x}{1+x^2}\,dx = \frac{1}{2}\int_{w=1}^{w=2} \frac{1}{w}\,dw = \frac{1}{2}\ln|w|\Big|_1^2 = \frac{1}{2}\ln 2.$$

   (b) We substitute $w = \cos x$, $dw = -\sin x\,dx$.

$$\int_{x=0}^{x=\frac{\pi}{4}} \frac{\sin x}{\cos x}\,dx = -\int_{w=1}^{w=\sqrt{2}/2} \frac{1}{w}\,dw$$

$$= -\ln|w|\Big|_1^{\sqrt{2}/2} = -\ln\frac{\sqrt{2}}{2} = \frac{1}{2}\ln 2.$$

5. We substitute $w = t + 2$, so $dw = dt$.

$$\int_{t=-1}^{t=e-2} \frac{1}{t+2}\,dt = \int_{w=1}^{w=e} \frac{dw}{w} = \ln|w|\Big|_1^e = \ln e - \ln 1 = 1.$$

9.

$$\int_{-1}^{3} (x^3 + 5x)\,dx = \frac{x^4}{4}\Big|_{-1}^{3} + \frac{5x^2}{2}\Big|_{-1}^{3} = 40.$$

13. Substitute $w = 1 + x^2$, $dw = 2x\,dx$. Then $x\,dx = \frac{1}{2}\,dw$, and

$$\int_{x=0}^{x=1} x(1+x^2)^{20}\,dx = \frac{1}{2}\int_{w=1}^{w=2} w^{20}\,dw = \frac{w^{21}}{42}\Big|_1^2 = \frac{299593}{6} = 49932\frac{1}{6}.$$

17. Substitute $w = x^2 + 4$, $dw = 2x\,dx$. Then,

$$\int_{x=4}^{x=1} x\sqrt{x^2+4}\,dx - \frac{1}{2}\int_{w=20}^{w=5} w^{\frac{1}{2}}\,dw - \frac{1}{3}w^{\frac{3}{2}}\Big|_{20}^{5}$$

$$= \frac{1}{3}\left(5^{\frac{3}{2}} - 8\cdot 5^{\frac{3}{2}}\right) = -\frac{7}{3}\cdot 5^{\frac{3}{2}} = -\frac{7}{3}\sqrt{125} = -26.087.$$

21. $f(t) = \sin \frac{1}{t}$ has no elementary antiderivative, so we will have to use left and right sums. With $n = 100$, left sum $= 0.5462$ and right sum $= 0.5582$. However, since f is not monotonic on $[1/4, 1]$ (see figure), we cannot be sure that the integral is between these values.

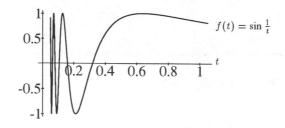

To get an upper and lower bound for this integral, divide the interval $[1/4, 1]$ into subintervals in such a way that f is monotonic on each one. Since $f'(t) = -\frac{1}{t^2} \cos \frac{1}{t} = 0$ when $\frac{1}{t} = \frac{\pi}{2}$ or $t = \frac{2}{\pi}$ (and this is the only point in $[1/4, 1]$ where $f'(t) = 0$), we will write

$$\int_{1/4}^{1} \sin \frac{1}{t} \, dt = \int_{1/4}^{2/\pi} \sin \frac{1}{t} \, dt + \int_{2/\pi}^{1} \sin \frac{1}{t} \, dt$$

Now, using $n = 100$, we find $0.209 < \int_{1/4}^{2/\pi} \sin \frac{1}{t} \, dt < 0.216$ (left sum $= 0.209$, right sum $= 0.216$, f is increasing), and $0.339 < \int_{2/\pi}^{1} \sin \frac{1}{t} \, dt < 0.340$ (here left sum $= 0.340$, right sum $= 0.339$ because f is decreasing). Thus

$$0.548 < \int_{1/4}^{1} \sin \frac{1}{t} \, dt < 0.556.$$

25. For the first integral, let $w = \sin x$, $dw = \cos x \, dx$. Then

$$\int e^{\sin x} \cos x \, dx = \int e^w \, dw.$$

For the second integral, let $w = \arcsin x$, $dw = \frac{1}{\sqrt{1-x^2}} dx$. Then

$$\int \frac{e^{\arcsin x}}{\sqrt{1 - x^2}} \, dx = \int e^w \, dw.$$

29. To find the area under the graph of $f(x) = xe^{x^2}$, we need to evaluate the definite integral

$$\int_0^2 xe^{x^2} \, dx.$$

This is done in Example 1, Section 7.3, using the substitution $w = x^2$, the result being

$$\int_0^2 xe^{x^2}\, dx = \frac{1}{2}(e^4 - 1) \approx 26.7991.$$

33. (a) Amount of water entering tank in a short period of time = rate×time = $r(t)\Delta t$.
    (b)

$$\text{Amount of water entering the tank between } t = 0 \text{ and } t = 5 \approx \sum_{i=0}^{n-1} r(t_i)\Delta t, \qquad \text{where } \Delta t = 5/n.$$

$$\text{Amount of water entering the tank between } t = 0 \text{ and } t = 5 = \int_0^5 r(t)\, dt.$$

(c) If $Q(t)$ is the amount of water in the tank at time t, then $Q'(t) = r(t)$. We want to calculate $Q(5) - Q(0)$. By the Fundamental Theorem,

$$\text{Amount which has entered tank} = Q(5) - Q(0) = \int_0^5 r(t)\, dt = \int_0^5 20e^{0.02t}\, dt = \frac{20}{0.02}e^{0.02t}\Big|_0^5$$

$$= 1000(e^{0.02(5)} - 1) \approx 105.17 \text{ gallons.}$$

(d) By the Fundamental Theorem again,

$$\text{Amount which has entered tank} = Q(t) - Q(0) = \int_0^t r(t)\, dt$$

$$Q(t) - 3000 = \int_0^t 20e^{0.02t}\, dt$$

so

$$Q(t) = 3000 + \int_0^t 20e^{0.02t}\, dt = 3000 + \frac{20}{0.02}e^{0.02t}\Big|_0^t$$

$$= 3000 + 1000(e^{0.02t} - 1)$$

$$= 1000e^{0.02t} + 2000.$$

7.4 SOLUTIONS

1. (a) i) $-x\sin x + \cos x$
       ii) $-2\sin 2x$
       iii) $-x^2\sin x + 2x\cos x$
       iv) $1 + \ln x$

(b)   i) Since $(x \ln x)' = 1 + \ln x$, it follows that $(x \ln x - x)' = 1 + \ln x - 1 = \ln x$. Therefore, $\int \ln x \, dx = x \ln x - x + C$.

ii) Since $(\cos 2x)' = -2 \sin 2x$, it follows that $(-\frac{1}{2} \cos 2x)' = \sin 2x$. Therefore, $\int \sin 2x \, dx = -\frac{1}{2} \cos 2x + C$.

iii) Since $(x \cos x)' = -x \sin x + \cos x$, it follows that $(-x \cos x + \sin x)' = x \sin x - \cos x + \cos x = x \sin x$. Therefore, $\int x \sin x \, dx = -x \cos x + \sin x + C$.

iv) Since $(x^2 \cos x)' = -x^2 \sin x + 2x \cos x$, and since $(2x \sin x)' = 2x \cos x + 2 \sin x$, it follows that

$$(-x^2 \cos x + 2x \sin x + 2 \cos x)' = x^2 \sin x - 2x \cos x + 2x \cos x + 2 \sin x - 2 \sin x$$
$$= x^2 \sin x.$$

Therefore, $\int x^2 \sin x \, dx = -x^2 \cos x + 2x \sin x + 2 \cos x + C$.

5.   Let $u = p$ and $v' = e^{(-0.1)p}$, $u' = 1$. Thus, $v = \int e^{(-0.1)p} \, dp = -10 e^{(-0.1)p}$. With this choice of u and v, integration by parts gives:

$$\int p e^{(-0.1)p} \, dp = p(-10 e^{(-0.1)p}) - \int (-10 e^{(-0.1)p}) \, dp$$
$$= -10 p e^{(-0.1)p} + 10 \int e^{(-0.1)p} \, dp$$
$$= -10 p e^{(-0.1)p} - 100 e^{(-0.1)p} + C.$$

9.   Let $u = t^2$, $v' = \sin t$ implying $v = -\cos t$ and $u' = 2t$. Integrating by parts, we get:

$$\int t^2 \sin t \, dt = -t^2 \cos t - \int 2t(-\cos t) \, dt.$$

Again, applying integration by parts with $u = t$, $v' = \cos t$, we have:

$$\int t \cos t \, dt = t \sin t + \cos t + C.$$

Thus

$$\int t^2 \sin t \, dt = -t^2 \cos t + 2t \sin t + 2 \cos t + C.$$

13.   Let $u = \theta + 1$ and $v' = \sin(\theta + 1)$, so $u' = 1$ and $v = -\cos(\theta + 1)$.

$$\int (\theta + 1) \sin(\theta + 1) \, d\theta = -(\theta + 1) \cos(\theta + 1) + \int \cos(\theta + 1) \, d\theta$$
$$= -(\theta + 1) \cos(\theta + 1) + \sin(\theta + 1) + C.$$

14. Let $u = \sin\theta$ and $v' = \sin\theta$, so $u' = \cos\theta$ and $v = -\cos\theta$. Then

$$\int \sin^2\theta \, d\theta = -\sin\theta\cos\theta + \int \cos^2\theta \, d\theta$$

$$= -\sin\theta\cos\theta + \int (1 - \sin^2\theta) \, d\theta$$

$$= -\sin\theta\cos\theta + \int 1 \, d\theta - \int \sin^2\theta \, d\theta.$$

By adding $\int \sin^2\theta \, d\theta$ to both sides of the above equation, we find that $2\int \sin^2\theta \, d\theta = -\sin\theta\cos\theta + \theta + C$, so $\int \sin^2\theta \, d\theta = -\frac{1}{2}\sin\theta\cos\theta + \frac{\theta}{2} + C'$.

17. Let $u = \ln x$, $v' = x^{-2}$. Then $v = -x^{-1}$ and $u' = x^{-1}$. Integrating by parts, we get:

$$\int x^{-2}\ln x \, dx = -x^{-1}\ln x - \int (-x^{-1}) \cdot x^{-1} \, dx$$

$$= -x^{-1}\ln x - x^{-1} + C.$$

21. $\displaystyle \int \frac{t+7}{\sqrt{5-t}} \, dt = \int \frac{t}{\sqrt{5-t}} \, dt + 7 \int (5-t)^{-1/2} \, dt.$

To calculate the first integral, we use integration by parts. Let $u = t$ and $v' = \frac{1}{\sqrt{5-t}}$, so $u' = 1$ and $v = -2(5-t)^{1/2}$. Then

$$\int \frac{t}{\sqrt{5-t}} \, dt = -2t(5-t)^{1/2} + 2\int (5-t)^{1/2} \, dt = -2t(5-t)^{1/2} - \frac{4}{3}(5-t)^{3/2} + C.$$

We can calculate the second integral directly: $7\displaystyle\int (5-t)^{-1/2} = -14(5-t)^{1/2} + C_1$. Thus

$$\int \frac{t+7}{\sqrt{5-t}} \, dt = -2t(5-t)^{1/2} - \frac{4}{3}(5-t)^{3/2} - 14(5-t)^{1/2} + C_2.$$

25. This integral can first be simplified by making the substitution $w = x^2$, $dw = 2x \, dx$. Then

$$\int x \arctan x^2 \, dx = \frac{1}{2}\int \arctan w \, dw.$$

To evaluate $\int \arctan w \, dw$, we'll use integration by parts. Let $u = \arctan w$ and $v' = 1$, so $u' = \frac{1}{1+w^2}$ and $v = w$. Then

$$\int \arctan w \, dw = w \arctan w - \int \frac{w}{1+w^2} \, dw = w\arctan w - \frac{1}{2}\ln|1+w^2| + C.$$

Since $1 + w^2$ is never negative, we can drop the absolute value signs. Thus, we have

$$\int x \arctan x^2 \, dx = \frac{1}{2} \left(x^2 \arctan x^2 - \frac{1}{2} \ln(1 + (x^2)^2) + C \right)$$
$$= \frac{1}{2} x^2 \arctan x^2 - \frac{1}{4} \ln(1 + x^4) + C.$$

29.   From integration by parts in Problem 14, we obtain

$$\int \sin^2 \theta \, d\theta = -\frac{1}{2} \sin \theta \cos \theta + \frac{1}{2} \theta + C.$$

Using the identity given in the book, we have

$$\int \sin^2 \theta \, d\theta = \int \frac{1 - \cos 2\theta}{2} \, d\theta = \frac{1}{2} \theta - \frac{1}{4} \sin 2\theta + C.$$

Although the answers differ in form, they are really the same, since (by one of the standard double angle formulas) $-\frac{1}{4} \sin 2\theta = -\frac{1}{4}(2 \sin \theta \cos \theta) = -\frac{1}{2} \sin \theta \cos \theta.$

31.   First, let $u = e^x$ and $v' = \sin x$, so $u' = e^x$ and $v = -\cos x$.
Thus $\int e^x \sin x \, dx = -e^x \cos x + \int e^x \cos x \, dx$. To calculate $\int e^x \cos x \, dx$, we again need to use integration by parts. Let $u = e^x$ and $v' = \cos x$, so $u' = e^x$ and $v = \sin x$.
Thus

$$\int e^x \cos x \, dx = e^x \sin x - \int e^x \sin x \, dx.$$

This gives

$$\int e^x \sin x \, dx = e^x \sin x - e^x \cos x - \int e^x \sin x \, dx.$$

By adding $\int e^x \sin x \, dx$ to both sides, we obtain

$$2 \int e^x \sin x \, dx = e^x (\sin x - \cos x) + C.$$

$$\text{Thus } \int e^x \sin x \, dx = \frac{1}{2} e^x (\sin x - \cos x) + C.$$

This problem could also be done in other ways; for example, we could have started with $u = \sin x$ and $v' = e^x$ as well.

32.   Let $u = e^\theta$ and $v' = \cos \theta$, so $u' = e^\theta$ and $v = \sin \theta$. Then $\int e^\theta \cos \theta \, d\theta = e^\theta \sin \theta - \int e^\theta \sin \theta \, d\theta.$

In Problem 31 we found that $\int e^x \sin x \, dx = \frac{1}{2} e^x (\sin x - \cos x) + C.$

$$\int e^\theta \cos \theta \, d\theta = e^\theta \sin \theta - \left[\frac{1}{2} e^\theta (\sin \theta - \cos \theta) \right] + C$$
$$= \frac{1}{2} e^\theta (\sin \theta + \cos \theta) + C.$$

33. We integrate by parts. Since in Problem 31 we found that $\int e^x \sin x \, dx = \frac{1}{2} e^x (\sin x - \cos x)$, we let $u = x$ and $v' = e^x \sin x$, so $u' = 1$ and $v = \frac{1}{2} e^x (\sin x - \cos x)$.

Then $\displaystyle\int x e^x \sin x \, dx = \frac{1}{2} x e^x (\sin x - \cos x) - \frac{1}{2} \int e^x (\sin x - \cos x) \, dx$

$\displaystyle = \frac{1}{2} x e^x (\sin x - \cos x) - \frac{1}{2} \int e^x \sin x \, dx + \frac{1}{2} \int e^x \cos x \, dx.$

Using Problems 31 and 32, we see that this equals

$$\frac{1}{2} x e^x (\sin x - \cos x) - \frac{1}{4} e^x (\sin x - \cos x) + \frac{1}{4} e^x (\sin x + \cos x) + C$$

$$= \frac{1}{2} x e^x (\sin x - \cos x) + \frac{1}{2} e^x \cos x + C.$$

37. We integrate by parts. Let $u = x^n$ and $v' = \cos ax$, so $u' = nx^{n-1}$ and $v = \frac{1}{a} \sin ax$. Then

$$\int x^n \cos ax \, dx = \frac{1}{a} x^n \sin ax - \int (nx^{n-1})(\frac{1}{a} \sin ax) \, dx$$

$$= \frac{1}{a} x^n \sin ax - \frac{n}{a} \int x^{n-1} \sin ax \, dx.$$

41. $\displaystyle\int_3^5 x \cos x \, dx = (\cos x + x \sin x) \Big|_3^5 = \cos 5 + 5 \sin 5 - \cos 3 - 3 \sin 3 \approx -3.944.$

44. We use integration by parts. Let $u = \arctan y$ and $v' = 1$, so $u' = \frac{1}{1+y^2}$ and $v = y$. Thus

$$\int_0^1 \arctan y \, dy = (\arctan y) y \Big|_0^1 - \int_0^1 \frac{y}{1+y^2} \, dy$$

$$-\frac{\pi}{4} - \frac{1}{2} \ln |1 + y^2| \Big|_0^1$$

$$= \frac{\pi}{4} - \frac{1}{2} \ln 2 \approx 0.439.$$

45. First we make the substitution $y = x^2$, so $dy = 2x \, dx$.
Thus

$$\int_{x=0}^{x=1} x \arctan x^2 \, dx = \frac{1}{2} \int_{y=0}^{y=1} \arctan y \, dy.$$

From Problem 44, we know that

$$\int_0^1 \arctan y \, dy = \frac{\pi}{4} - \frac{\ln 2}{2}.$$

Thus

$$\int_0^1 x \arctan x^2 \, dx = \frac{1}{2}(\frac{\pi}{4} - \frac{1}{2}\ln 2) \approx 0.219.$$

49.   (a)   One way to avoid integrating by parts is to take the derivative of the right hand side instead. Since $\int e^{ax} \sin bx \, dx$ is the antiderivative of $e^{ax} \sin bx$,

$$\begin{aligned}
e^{ax} \sin bx &= \frac{d}{dx}[e^{ax}(A \sin bx + B \cos bx) + C] \\
&= ae^{ax}(A \sin bx + B \cos bx) + e^{ax}(Ab \cos bx - Bb \sin bx) \\
&= e^{ax}[(aA - bB) \sin bx + (aB + bA) \cos bx].
\end{aligned}$$

Thus $aA - bB = 1$ and $aB + bA = 0$. Solving for A and B in terms of a and b, we get

$$A = \frac{a}{a^2 + b^2}, \quad B = -\frac{b}{a^2 + b^2}.$$

Thus

$$\int e^{ax} \sin bx = e^{ax}\left(\frac{a}{a^2 + b^2} \sin bx - \frac{b}{a^2 + b^2} \cos bx\right) + C.$$

(b)   If we go through the same process, we find

$$ae^{ax}[(aA - bB) \sin bx + (aB + bA) \cos bx] = e^{ax} \cos bx.$$

Thus $aA - bB = 0$, and $aB + bA = 1$. In this case, solving for A and B yields

$$A = \frac{b}{a^2 + b^2}, \quad B = \frac{a}{a^2 + b^2}.$$

Thus $\int e^{ax} \cos bx = e^{ax}(\frac{b}{a^2+b^2} \sin bx + \frac{a}{a^2+b^2} \cos bx) + C.$

7.5   SOLUTIONS

1.   See the solutions to Problems 5– 13. Answers may vary, as there may be more than one way to approach a problem.

5. $\left(\dfrac{1}{2}x^3 - \dfrac{3}{4}x^2 + \dfrac{3}{4}x - \dfrac{3}{8}\right)e^{2x} + C.$

        (Let $a = 2, p(x) = x^3$ in III- 14.)

9. $\dfrac{1}{6}x^6 \ln x - \dfrac{1}{36}x^6 + C.$ (Let $n = 5$ in III- 13.)

13. $\dfrac{1}{\sqrt{3}} \arctan \dfrac{y}{\sqrt{3}} + C.$

      (Let $a = \sqrt{3}$ in V- 24).

17.

$$\int y^2 \sin 2y \, dy = -\frac{1}{2}y^2 \cos 2y + \frac{1}{4}(2y)\sin 2y + \frac{1}{8}(2)\cos 2y + C$$

$$= -\frac{1}{2}y^2 \cos 2y + \frac{1}{2}y \sin 2y + \frac{1}{4}\cos 2y + C.$$

(Use $a = 2, p(y) = y^2$ in III- 15.)

21. If we make the substitution $w = 2z^2$ then $dw = 4z \, dz$, and the integral becomes:

$$\int ze^{2z^2}\cos(2z^2) \, dz = \frac{1}{4}\int e^w \cos w \, dw$$

Now we can use Formula 9 from the table of integrals to get:

$$\frac{1}{4}\int e^w \cos w \, dw = \frac{1}{4}\left[\frac{1}{2}e^w(\cos w + \sin w) + C\right]$$

$$= \frac{1}{8}e^w(\cos w + \sin w) + C$$

$$= \frac{1}{8}e^{2z^2}(\cos 2z^2 + \sin 2z^2) + C$$

25. Substitute $w = z^2$, $dw = 2z \, dz$. Using IV- 17,

$$\int z \sin^3(z^2) \, dz = \frac{1}{2}\int \sin^3 w \, dw = \frac{1}{2}[-\frac{1}{3}\sin^2 w \cos w + \frac{2}{3}\int \sin w \, dw]$$

$$= -\frac{1}{6}\sin^2 w \cos w - \frac{1}{3}\cos w + C$$

$$= -\frac{1}{6}\sin^2(z^2)\cos(z^2) - \frac{1}{3}\cos(z^2) + C.$$

29. Substitute $w = 2\theta$, $dw = 2 \, d\theta$. Then use IV 19, letting $m = 2$.

$$\int \frac{1}{\sin^2 2\theta} \, d\theta = \frac{1}{2}\int \frac{1}{\sin^2 w} \, dw = \frac{1}{2}(-\frac{\cos w}{\sin w}) + C = -\frac{1}{2\tan w} + C = -\frac{1}{2\tan 2\theta} + C.$$

33. $-\dfrac{1}{4}(9 - 4x^2)^{\frac{1}{2}} + C.$

(Substitute $w = 9 - 4x^2$, $dw = -8x\,dx$. You need not use the table.)

37.

$$\int \frac{dz}{z(z - 3)} = -\frac{1}{3}(\ln|z| - \ln|z - 3|) + C.$$

(Let $a = 0, b = 3$ in V- 26.)

41.

$$\arcsin \frac{x + 1}{\sqrt{2}} + C.$$

(Substitute $w = x + 1$, and then apply VI- 28 with $a = \sqrt{2}$).

45. Using long division, we find that

$$\frac{x^3 + 3}{x^2 - 3x + 2} = x + 3 + \frac{7x - 3}{x^2 - 3x + 2}.$$

Thus

$$\int \frac{x^3 + 3}{x^2 - 3x + 2}\,dx = \int \left(x + 3 + \frac{7x - 3}{x^2 - 3x + 2} \right)\,dx$$

$$= \int (x + 3)\,dx + \int \frac{7x - 3}{(x - 1)(x - 2)}\,dx.$$

Using V- 27 (with $a = 1, b = 2, c = 7$, and $d = -3$) we have

$$\int \frac{7x - 3}{(x - 1)(x - 2)}\,dx = -4\ln|x - 1| + 11\ln|x - 2| + C.$$

Thus

$$\int \frac{x^3 + 3}{x^2 - 3x + 2}\,dx = \frac{x^2}{2} + 3x - 4\ln|x - 1| + 11\ln|x - 2| + C.$$

49.

$$\int \frac{5z - 13}{z^2 - 5z + 6}\,dz = \int \frac{5z - 13}{(z - 3)(z - 2)}\,dz.$$

Let $a = 3, b = 2, c = 5$, and $d = -13$ in V- 27.

$$\int \frac{5z - 13}{(z - 3)(z - 2)}\,dz = 2\ln|z - 3| + 3\ln|z - 2| + C.$$

53.

$$\int_0^1 \frac{dx}{x^2 + 2x + 5} == \int_0^1 \frac{dx}{(x+1)^2 + 4}$$

$$= \frac{1}{2} \arctan \frac{x+1}{2} \Big|_0^1 = \frac{1}{2} \arctan 1 - \frac{1}{2} \arctan \frac{1}{2} \approx 0.1609.$$

(Substitute $w = x + 1$ and use V- 24).

$\dfrac{1}{x^2 + 2x + 5}$ is monotonic over $0 \le x \le 1$, so we expect the value of the integral to be between the left- and right-hand sums. Using 100 subintervals, we find

$$0.1605 < \int_0^1 \frac{dx}{x^2 + 2x + 5} < 0.1613$$

which matches our result.

57. Use $\sin 2\theta = 2 \sin \theta \cos \theta$. Then

$$\int_0^2 \cos \theta \sin 2\theta \, d\theta = 2 \int_0^2 \cos^2 \theta \sin \theta \, d\theta = (-2) \frac{\cos^3 \theta}{3} \Big|_0^2 \approx 0.7147.$$

$\cos \theta \sin 2\theta$ is not monotonic over the interval $0 \le \theta \le 2$, but we find using 100 subintervals that the left-hand sum ≈ 0.7115, the right-hand sum ≈ 0.7178, and the average ≈ 0.7147, so these results match our answer very well.

61. $\displaystyle\int_0^1 \sqrt{3 - x^2} \, dx = (\frac{1}{2} x \sqrt{3 - x^2} + \frac{3}{2} \arcsin \frac{x}{\sqrt{3}}) \Big|_0^1 \approx 1.630.$

(Let $a = \sqrt{3}$ in VI- 30 and VI- 28).

Again, $\sqrt{3 - x^2}$ is monotonic on $0 \le x \le 1$, and using 100 subintervals, we find $1.629 < \int_0^1 \sqrt{3 - x^2} \, dx < 1.632$, which matches our result.

65. (a) Since $R(T)$ is the rate or production, we find the total production by integrating:

$$\int_0^N R(t) \, dt = \int_0^N (A + Be^{-t} \sin(2\pi t)) \, dt$$

$$= NA + B \int_0^N e^{-t} \sin(2\pi t) \, dt.$$

Let $a = -1$ and $b = 2\pi$ in II- 8.

$$= NA + \frac{B}{1 + 4\pi^2} e^{-t} (-\sin(2\pi t) - 2\pi \cos(2\pi t)) \Big|_0^N .$$

Since N is an integer (so $\sin 2\pi N = 0$ and $\cos 2\pi N = 1$),

$$\int_0^N R(t)\, dt = NA + B\frac{2\pi}{1+4\pi^2}(1 - e^{-N}).$$

Thus the total production is $NA + \frac{2\pi B}{1+4\pi^2}(1 - e^{-N})$ over the first N years.

(b) The average production over the first N years is

$$\int_0^N \frac{R(t)\, dt}{N} = A + \frac{2\pi B}{1+4\pi^2}\left(\frac{1 - e^{-N}}{N}\right).$$

(c) As $N \to \infty$, $A + \frac{2\pi B}{1+4\pi^2}\frac{1-e^{-N}}{N} \to A$, since the second term in the sum goes to 0. This is why A is called the average!

(d) When t gets large, the term $Be^{-t}\sin(2\pi t)$ gets very small. Thus, $R(t) \approx A$ for most t, so it makes sense that the average of $\int_0^N R(t)\, dt$ is A as $N \to \infty$.

(e) This model is not reasonable for long periods of time, since an oil well has finite capacity and will eventually "run dry." Thus, we cannot expect average production to be close to constant over a long period of time.

69. (a) Let $y = a\ln x + b\ln(1+x) + \dfrac{c}{1+x}$. Then

$$\frac{dy}{dx} = \frac{a}{x} + \frac{b}{1+x} - \frac{c}{(1+x)^2} = \frac{a(1+x)^2 + bx(1+x) - cx}{x(1+x)^2}.$$

(b) The denominator of $\frac{dy}{dx}$ is the same as that of the integrand. The numerator of $\frac{dy}{dx}$ is a polynomial in x of the same degree as that of the integrand:

$$\begin{aligned}
\text{Numerator} &= a(1+x)^2 + bx(1+x) - cx \\
&= ax^2 + 2ax + a + bx^2 + bx - cx \\
&= (a+b)x^2 + (2a + b - c)x + a
\end{aligned}$$

Set the numerator $= 1 + x^2$:

$$1 + x^2 = (a+b)x^2 + (2a + b - c)x + a.$$

Then the coefficients of the polynomial are given by

$$\begin{aligned}
a + b &= 1, \\
2a + b - c &= 0, \\
a &= 1
\end{aligned}$$

which implies that $a = 1$, $b = 0$, and $c = 2$. So, plugging in the values of a, b, and c to the equation

$$y = a\ln x + b\ln(1+x) + \frac{c}{1+x},$$

we get

$$\ln x + \frac{2}{1+x} = \int \frac{1+x^2}{x(1+x)^2} \, dx.$$

Thus

$$\int_1^2 \frac{1+x^2}{x(1+x)^2} \, dx = \left(\ln x + \frac{2}{1+x} \right) \Big|_1^2 = \ln 2 - \frac{1}{3}.$$

7.6   SOLUTIONS

1.

TABLE 7.1

n	1	2	4
LEFT	40	40.7846	41.7116
RIGHT	51.2250	46.3971	44.5179
TRAP	45.6125	43.5909	43.1147
MID	41.5692	42.6386	42.8795

5.

TABLE 7.2

N	Left Sum	Right Sum	Trapezoid	Midpoint
10	0.14047	0.18842	0.16445	0.16335
100	0.16132	0.16612	0.16372	0.16371
1000	0.16347	0.16395	0.16371	0.16371

The function is increasing on the interval $0 \le \theta \le 1$, so left sums give an underestimate and right sums an overestimate. It is concave down on all of the interval, so the trapezoid sum gives an underestimate and the midpoint sum gives an overestimate.

9. (a) (i) $\text{LEFT}(32) = 13.6961, \text{RIGHT}(32) = 14.3437, \text{TRAP}(32) = 14.0199$

Exact value $= (x \ln x - x) \Big|_1^{10} \approx 14.02585093$

(ii) $\text{LEFT}(32) = 50.3180, \text{RIGHT}(32) = 57.0178, \text{TRAP}(32) = 53.6679$

Exact value $= e^x \Big|_0^4 \approx 53.59815003$

(b) Both $\ln x$ and e^x are increasing, so the left sum underestimates and the right sum overestimates.

(i) $\text{LEFT}(32) \le \text{TRAP}(32) \le$ Actual value $\le \text{RIGHT}(32)$

(ii) $\text{LEFT}(32) \le$ Actual value $\le \text{TRAP}(32) \le \text{RIGHT}(32)$

The trapezoid rule is an overestimate if f is concave up, and an underestimate if it is concave down.

Since $\ln x$ is concave down, the trapezoidal estimate is too small. Since e^x is concave up, the trapezoidal estimate is too large. In each case, however, the trapezoidal estimate should be better than the left- or right-hand sums, since it is the average of the two.

13. (a) $\displaystyle\int_0^{2\pi} \sin\theta \, d\theta = -\cos\theta \Big|_0^{2\pi} = 0.$

   (b) MID(1) is 0 since the midpoint of 0 and 2π is π, and $\sin\pi = 0$. Thus MID(1) $= 2\pi(\sin\pi) = 0$.
        MID(2) is 0 since the midpoints we use are $\pi/2$ and $3\pi/2$, and $\sin(\pi/2) = -\sin(3\pi/2)$. So MID(2) $= \pi\sin(\pi/2) + \pi\sin(3\pi/2) = 0$.

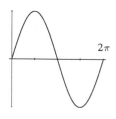

   (c) MID(3) $= 0$.
        In general, MID(n) $= 0$ for all n, even though your calculator (because of round-off error) might not return it as such. The reason is that $\sin(x) = -\sin(2\pi - x)$. If we use MID($n$), we will always take sums where we are adding pairs of the form $\sin(x)$ and $\sin(2\pi - x)$, so the sum will cancel to 0. (If n is odd, we will get a $\sin\pi$ in the sum which doesn't pair up with anything — but $\sin\pi$ is already 0!)

17.

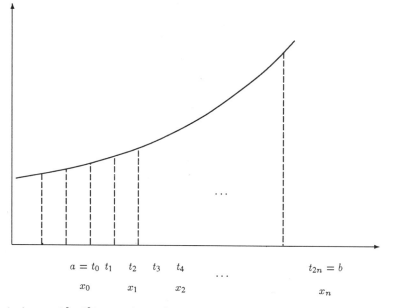

Divide the interval $[a, b]$ into n pieces, by $x_0, x_1, x_2, \ldots, x_n$, and also into $2n$ pieces, by $t_0, t_1, t_2, \ldots, t_{2n}$. Then the x's coincide with the even t's, so $x_0 = t_0$, $x_1 = t_2$, $x_2 = t_4$, $\ldots$, $x_n = t_{2n}$ and $\Delta t = \frac{1}{2}\Delta x$.

$$\text{LEFT}(n) = f(x_0)\Delta x + f(x_1)\Delta x + \cdots + f(x_{n-1})\Delta x$$

Since MID(n) is obtained by evaluating f at the midpoints $t_1, t_3, t_5, \ldots$ of the x intervals, we get

$$\text{MID}(n) = f(t_1)\Delta x + f(t_3)\Delta x + \cdots + f(t_{2n-1})\Delta x$$

Now

$$\text{LEFT}(2n) = f(t_0)\Delta t + f(t_1)\Delta t + f(t_2)\Delta t + \cdots + f(t_{2n-1})\Delta t.$$

Regroup terms, putting all the even t's first, the odd t's last:

$$\text{LEFT}(2n) = f(t_0)\Delta t + f(t_2)\Delta t + \cdots + f(t_{2n-2})\Delta t + f(t_1)\Delta t + f(t_3)\Delta t + \cdots + f(t_{2n-1})\Delta t$$

$$= \underbrace{f(x_0)\frac{\Delta x}{2} + f(x_1)\frac{\Delta x}{2} + \cdots + f(x_{n-1})\frac{\Delta x}{2}}_{\text{LEFT}(n)/2} + \underbrace{f(t_1)\frac{\Delta x}{2} + f(t_3)\frac{\Delta x}{2} + \cdots + f(t_{2n-1})\frac{\Delta x}{2}}_{\text{MID}(n)/2}$$

So

$$\text{LEFT}(2n) = \frac{1}{2}(\text{LEFT}(n) + \text{MID}(n))$$

21.   Approximate $\int_1^{100001} \frac{1}{x}\, dx$, by rectangles, using $n = 100{,}000$ so $\Delta x = 1$.

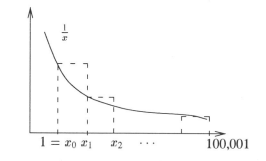

Then

$$\text{LEFT}(100{,}000) = f(1) \cdot 1 + f(2) \cdot 1 + \cdots + f(100{,}000) \cdot 1$$

$$= \frac{1}{1} + \frac{1}{2} + \cdots + \frac{1}{100{,}000} = \sum_{k=1}^{100{,}000} \frac{1}{k}$$

Since the left sum is an overestimate,

$$\int_1^{100001} \frac{1}{x}\, dx < \text{LEFT}(100{,}000),$$

and since

$$\int_1^{100001} \frac{1}{x}\, dx = \ln(100{,}001) - \ln 1 = \ln(100{,}001),$$

so

$$\ln 100001 < \sum_{k=1}^{100000} \frac{1}{k}.$$

Now imagine all the rectangles moved one unit to the left; they are the right sum approximation to

$$\int_1^{100000} \frac{1}{x}\, dx + \text{area of first rectangle}$$

and this time they give an underestimate.

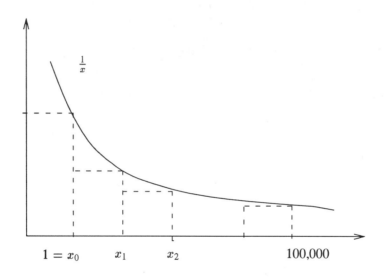

The area of the rectangles is our sum, so

$$\sum_{k=1}^{100000} \frac{1}{k} < \int_1^{100000} \frac{1}{x}\, dx + \text{area of first rectangle}$$

So

$$\sum_{k=1}^{100000} \frac{1}{k} < (\ln 100000) + 1$$

Thus

$$\ln(100001) < \sum_{k=1}^{100000} \frac{1}{k} < (\ln 100{,}000) + 1$$

$$11.5 < \sum_{k=1}^{100000} \frac{1}{k} < 12.5 \quad \text{so} \quad \sum_{k=1}^{100000} \frac{1}{k} \approx 12.$$

7.7 SOLUTIONS

1.

$$\text{SIMP} = \frac{1}{3}(2\,\text{MID} + \text{TRAP}) = \frac{1}{3}\left(2\,\text{MID} + \frac{\text{LEFT} + \text{RIGHT}}{2}\right)$$
$$= \frac{1}{3}\left(2 \cdot 0.6857 + \frac{0.8333 + 0.5833}{2}\right)$$
$$\approx 0.6932$$

The actual value is $\ln 2 \approx 0.6932$.

5. 1.5699. ($n = 10$ intervals, or more)

9. 0.904524. ($n = 10$ intervals, or more)

13. (a) For the left-hand rule, error is approximately proportional to $\frac{1}{n}$. If we let n_p be the number of subdivisions needed for accuracy to p places, then there is a constant k such that

$$5 \times 10^{-5} = \frac{1}{2} \times 10^{-4} \approx \frac{k}{n_4}$$

$$5 \times 10^{-9} = \frac{1}{2} \times 10^{-8} \approx \frac{k}{n_8}$$

$$5 \times 10^{-13} = \frac{1}{2} \times 10^{-12} \approx \frac{k}{n_{12}}$$

$$5 \times 10^{-21} = \frac{1}{2} \times 10^{-20} \approx \frac{k}{n_{20}}$$

Thus the ratios $n_4 : n_8 : n_{12} : n_{20} \approx 1 : 10^4 : 10^8 : 10^{16}$, and assuming the computer time necessary is proportional to n_p, the computer times are approximately

4 places:	2 seconds	
8 places:	2×10^4 seconds	≈ 7 hours
12 places:	2×10^8 seconds	≈ 6 years
20 places:	2×10^{16} seconds	≈ 600 million years

(b) For the trapezoidal rule, error is approximately proportional to $\frac{1}{n^2}$. If we let N_p be the number of subdivisions needed for accuracy to p places, then there is a constant C such that

$$5 \times 10^{-5} = \frac{1}{2} \times 10^{-4} \approx \frac{C}{N_4^{2}}$$

$$5 \times 10^{-9} = \frac{1}{2} \times 10^{-8} \approx \frac{C}{N_8{}^2}$$

$$5 \times 10^{-13} = \frac{1}{2} \times 10^{-12} \approx \frac{C}{N_{12}{}^2}$$

$$5 \times 10^{-21} = \frac{1}{2} \times 10^{-20} \approx \frac{C}{N_{20}{}^2}$$

Thus the ratios $N_4{}^2 : N_8{}^2 : N_{12}{}^2 : N_{20}{}^2 \approx 1 : 10^4 : 10^8 : 10^{16}$, and the ratios $N_4 : N_8 : N_{12} : N_{20} \approx 1 : 10^2 : 10^4 : 10^8$. So the computer times are approximately

4 places:	2 seconds	
8 places:	2×10^2 seconds	$\approx$ 3 minutes
12 places:	2×10^4 seconds	$\approx$ 7 hours
20 places:	2×10^8 seconds	$\approx$ 6 years

17.   False. If the function $f(x)$ is a line, then the trapezoid rule gives the exact answer to $\int_a^b f(x)\,dx$.

20.   (a)   If $f(x) = 1$, then

$$\int_a^b f(x)\,dx = (b - a).$$

Also,

$$\frac{h}{3}\left(\frac{f(a)}{2} + 2f(m) + \frac{f(b)}{2}\right) = \frac{b-a}{3}\left(\frac{1}{2} + 2 + \frac{1}{2}\right) = (b - a).$$

So the equation holds for $f(x) = 1$.
   If $f(x) = x$, then

$$\int_a^b f(x)\,dx = \left.\frac{x^2}{2}\right|_a^b = \frac{b^2 - a^2}{2}.$$

Also,

$$\frac{h}{3}\left(\frac{f(a)}{2} + 2f(m) + \frac{f(b)}{2}\right) = \frac{b-a}{3}\left(\frac{a}{2} + 2\frac{a+b}{2} + \frac{b}{2}\right)$$

$$= \frac{b-a}{3}\left(\frac{a}{2} + a + b + \frac{b}{2}\right)$$

$$= \frac{b-a}{3}\left(\frac{3}{2}b + \frac{3}{2}a\right)$$

$$= \frac{(b-a)(b+a)}{2}$$

$$= \frac{b^2 - a^2}{2}.$$

So the equation holds for $f(x) = x$.

If $f(x) = x^2$, then $\int_a^b f(x)\,dx = \left.\frac{x^3}{3}\right|_a^b = \frac{b^3 - a^3}{3}$. Also,

$$\frac{h}{3}\left(\frac{f(a)}{2} + 2f(m) + \frac{f(b)}{2}\right) = \frac{b-a}{3}\left(\frac{a^2}{2} + 2\left(\frac{a+b}{2}\right)^2 + \frac{b^2}{2}\right)$$

$$= \frac{b-a}{3}\left(\frac{a^2}{2} + \frac{a^2 + 2ab + b^2}{2} + \frac{b^2}{2}\right)$$

$$= \frac{b-a}{3}\left(\frac{2a^2 + 2ab + 2b^2}{2}\right)$$

$$= \frac{b-a}{3}\left(a^2 + ab + b^2\right)$$

$$= \frac{b^3 - a^3}{3}.$$

So the equation holds for $f(x) = x^2$.

(b) For any quadratic function, $f(x) = Ax^2 + Bx + C$, the "Facts about Sums and Constant Multiples of Integrands" give us:

$$\int_a^b f(x)\,dx = \int_a^b (Ax^2 + Bx + C)\,dx = A\int_a^b x^2\,dx + B\int_a^b x\,dx + C\int_a^b 1\,dx.$$

Now we use the results of part (a) to get:

$$\int_a^b f(x)\,dx = A\frac{h}{3}\left(\frac{a^2}{2} + 2m^2 + \frac{b^2}{2}\right) + B\frac{h}{3}\left(\frac{a}{2} + 2m + \frac{b}{2}\right) + C\frac{h}{3}\left(\frac{1}{2} + 2\cdot 1 + \frac{1}{2}\right)$$

$$= \frac{h}{3}\left(\frac{Aa^2 + Ba + C}{2} + 2(Am^2 + Bm + C) + \frac{Ab^2 + Bb + C}{2}\right)$$

$$= \frac{h}{3}\left(\frac{f(a)}{2} + 2f(m) + \frac{f(b)}{2}\right).$$

21. (a) Suppose $q_i(x)$ is the quadratic function approximating $f(x)$ on the subinterval $[x_i, x_{i+1}]$, and m_i is the midpoint of the interval, $m_i = (x_i + x_{i+1})/2$. Then, using the equation in Problem 20, with $a = x_i$ and $b = x_{i+1}$ and $h = \Delta x = x_{i+1} - x_i$:

$$\int_{x_i}^{x_{i+1}} f(x)\,dx \approx \int_{x_i}^{x_{i+1}} q_i(x)\,dx = \frac{\Delta x}{3}\left(\frac{q_i(x_i)}{2} + 2q_i(m_i) + \frac{q_i(x_{i+1})}{2}\right).$$

(b) Summing over all subintervals gives

$$\int_a^b f(x)\,dx \approx \sum_{i=0}^{n-1}\int_{x_i}^{x_{i+1}} q_i(x)\,dx = \sum_{i=0}^{n-1}\frac{\Delta x}{3}\left(\frac{q_i(x_i)}{2} + 2q_i(m_i) + \frac{q_i(x_{i+1})}{2}\right).$$

Splitting the sum into two parts:

$$= \frac{2}{3} \sum_{i=0}^{n-1} q_i(m_i)\Delta x + \frac{1}{3} \sum_{i=0}^{n-1} \frac{q_i(x_i) + q_i(x_{i+1})}{2} \Delta x$$

$$= \frac{2}{3} \text{MID}(n) + \frac{1}{3} \text{TRAP}(n)$$

$$= \text{SIMP}(n).$$

7.8 SOLUTIONS

1.

$$\int_1^\infty e^{-2x}\, dx = \lim_{b \to \infty} \int_1^b e^{-2x}\, dx = \lim_{b \to \infty} \left. -\frac{e^{-2x}}{2} \right|_1^b$$

$$= \lim_{b \to \infty} (-e^{-2b}/2 + e^{-2}/2) = 0 + e^{-2}/2 = e^{-2}/2,$$

where the first limit is 0 because $\lim_{x \to \infty} e^{-x} = 0$.

5.

$$\int_\pi^\infty \sin y\, dy = \lim_{b \to \infty} \int_\pi^b \sin y\, dy$$

$$= \lim_{b \to \infty} \left. (-\cos y) \right|_\pi^b$$

$$= \lim_{b \to \infty} [-\cos b - (-\cos \pi)].$$

As $b \to \infty$, $-\cos b$ fluctuates between -1 and 1, so the limit fails to exist: the integral diverges. (This doesn't follow right from the fact that $\sin y$ fluctuates between -1 and 1!)

9. This integral is improper because $1/v$ blows up at $v = 0$. To evaluate it, we must split the region of integration up into two pieces, from 0 to 1 and from -1 to 0. But notice,

$$\int_0^1 \frac{1}{v}\, dv = \lim_{b \to 0^+} \int_b^1 \frac{1}{v}\, dv = \lim_{b \to 0^+} \left(\left. \ln v \right|_b^1 \right) = -\ln b.$$

As $b \to 0^+$, this goes to infinity and the integral diverges, so our original integral also diverges.

13.

$$\int_1^\infty \frac{y}{y^4 + 1}\, dy = \lim_{b \to \infty} \frac{1}{2} \int_1^b \frac{2y}{(y^2)^2 + 1}\, dy$$

$$= \lim_{b \to \infty} \left. \frac{1}{2} \arctan(y^2) \right|_1^b$$

$$= \lim_{b \to \infty} \frac{1}{2}[\arctan(b^2) - \arctan 1]$$
$$= (1/2)[\pi/2 - \pi/4] = \pi/8.$$

17. With the substitution $w = \ln x$, $dw = \frac{1}{x}dx$,

$$\int \frac{dx}{x \ln x} = \int \frac{1}{w} \, dw = \ln|w| + C = \ln|\ln x| + C$$

so

$$\int_2^\infty \frac{dx}{x \ln x} = \lim_{b \to \infty} \int_2^b \frac{dx}{x \ln x}$$
$$= \lim_{b \to \infty} \ln|\ln x| \Big|_2^b$$
$$= \lim_{b \to \infty} [\ln|\ln b| - \ln|\ln 2|].$$

As $b \to \infty$, the limit goes to ∞ and hence the integral diverges.

21. As in Problem 17, $\int \frac{dx}{x \ln x} = \ln|\ln x| + C$, so

$$\int_1^2 \frac{dx}{x \ln x} = \lim_{b \to 1+} \int_b^2 \frac{dx}{x \ln x}$$
$$= \lim_{b \to 1+} \ln|\ln x| \Big|_b^2$$
$$= \lim_{b \to 1+} \ln(\ln 2) - \ln(\ln b).$$

As $b \to 1^+$, $\ln(\ln b) \to -\infty$, so the integral diverges.

25. The curve has an asymptote at $t = \frac{\pi}{2}$, and so the area integral is improper there.

$$\text{Area} = \int_0^{\frac{\pi}{2}} \frac{dt}{\cos^2 t} = \lim_{b \to \frac{\pi}{2}} \int_0^b \frac{dt}{\cos^2 t} = \lim_{b \to \frac{\pi}{2}} \tan t \Big|_0^b,$$

which diverges. Therefore the area is infinite.

29. The energy required is

$$E = \int_1^\infty \frac{kq_1 q_2}{r^2} \, dr = kq_1 q_2 \lim_{b \to \infty} -\frac{1}{r} \Big|_1^b$$
$$= (9 \times 10^9)(1)(1)(1) = 9 \times 10^9 \text{ joules}$$

7.9 SOLUTIONS

1. It converges:

$$\int_{50}^{\infty} \frac{dz}{z^3} = \lim_{b \to \infty} \int_{50}^{b} \frac{dz}{z^3} = \lim_{b \to \infty} \left(-\frac{1}{2} z^{-2} \Big|_{50}^{b} \right) = \frac{1}{2} \lim_{b \to \infty} \left(\frac{1}{50^2} - \frac{1}{b^2} \right) = \frac{1}{5000}$$

5. Since $\dfrac{1}{1+x} \geq \dfrac{1}{2x}$ and $\dfrac{1}{2} \displaystyle\int_{0}^{\infty} \dfrac{1}{x}\, dx$ diverges, we have that $\displaystyle\int_{1}^{\infty} \dfrac{dx}{1+x}$ diverges.

9. Since $\dfrac{1}{e^z + 2^z} < \dfrac{1}{e^z} = e^{-z}$ for $z \geq 0$, and $\displaystyle\int_{0}^{\infty} e^{-z}\, dz$ converges, $\displaystyle\int_{0}^{\infty} \dfrac{dz}{e^z + 2^z}$ converges.

13. Since $\dfrac{3 + \sin \alpha}{\alpha} \geq \dfrac{2}{\alpha}$ for $\alpha \geq 4$, and since $\displaystyle\int_{4}^{\infty} \dfrac{2}{\alpha}\, d\alpha$ diverges, then $\displaystyle\int_{4}^{\infty} \dfrac{3 + \sin \alpha}{\alpha}\, d\alpha$ diverges.

17. If we integrate e^{-x^2} from 1 to 10, we get 0.139. This answer doesn't change noticeably if you extend the region of integration to from 1 to 11, say, or even up to 1000. There's a reason for this; and the reason is that the tail, $\int_{10}^{\infty} e^{-x^2}\, dx$, is very small indeed. In fact

$$\int_{10}^{\infty} e^{-x^2}\, dx \leq \int_{10}^{\infty} e^{-x}\, dx = e^{-10},$$

which is very small. (In fact, the tail integral is less than $e^{-100}/10$. Can you prove that? [Hint: $e^{-x^2} \leq e^{-10x}$ for $x \geq 10$.])

21. (a) Since $e^{-x^2} \leq e^{-3x}$ for $x \geq 3$,

$$\int_{3}^{\infty} e^{-x^2}\, dx \leq \int_{3}^{\infty} e^{-3x}\, dx$$

Now

$$\int_{3}^{\infty} e^{-3x}\, dx = \lim_{b \to \infty} \int_{3}^{b} e^{-3x}\, dx = \lim_{b \to \infty} -\frac{1}{3} e^{-3x} \Big|_{3}^{b}$$

$$= \lim_{b \to \infty} \frac{e^{-9}}{3} - \frac{e^{-3b}}{3} = \frac{e^{-9}}{3}.$$

Thus

$$\int_{3}^{\infty} e^{-x^2}\, dx \leq \frac{e^{-9}}{3}.$$

(b) By reasoning similar to part (a),

$$\int_{n}^{\infty} e^{-x^2}\, dx \leq \int_{n}^{\infty} e^{-nx}\, dx,$$

and

$$\int_n^\infty e^{-nx}\,dx = \frac{1}{n}e^{-n^2},$$

so

$$\int_n^\infty e^{-x^2}\,dx \le \frac{1}{n}e^{-n^2}.$$

25. (a) The tangent line to e^t has slope $(e^t)' = e^t$. Thus at $t = 0$, the slope is $e^0 = 1$. The line passes through $(0, e^0) = (0, 1)$. Thus the equation of the tangent line is $y = 1+t$. Since e^t is everywhere concave up, its graph is always above the graph of any of its tangent lines; in particular, e^t is always above the line $y = 1 + t$. This is tantamount to saying

$$1 + t \le e^t,$$

with equality holding only at the point of tangency, $t = 0$.

(b) If $t = \dfrac{1}{x}$, then the above inequality becomes

$$1 + \frac{1}{x} \le e^{1/x}, \text{ or } e^{1/x} - 1 \ge \frac{1}{x}.$$

Since $t = \dfrac{1}{x}$, t is never zero. Therefore, the inequality is strict, and we write

$$e^{1/x} - 1 > \frac{1}{x}.$$

(c) Since $e^{1/x} - 1 > \dfrac{1}{x}$,

$$\frac{1}{x^5\left(e^{1/x} - 1\right)} < \frac{1}{x^5\left(\frac{1}{x}\right)} = \frac{1}{x^4}.$$

Since $\displaystyle\int_1^\infty \frac{dx}{x^4}$ converges, $\displaystyle\int_1^\infty \frac{dx}{x^5\left(e^{1/x} - 1\right)}$ converges.

7.10 SOLUTIONS

1.

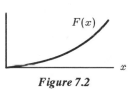

Figure 7.2

By the Fundamental Theorem of Calculus, $f(x) = F'(x)$. Since f is positive and increasing, F is increasing and concave up. Notice that since $F(0) = \int_0^0 f(t)dt = 0$, the graph of F must start from the origin.

5. (a) Again using 0.00001 as the lower limit, because the integral is improper, gives Si(4) = 1.76, Si(5) = 1.55.

   (b) Si(x) decreases when the integrand is negative, which occurs when $\pi < x < 2\pi$.

9. $(1+x)^{200}$.

13. $\dfrac{d}{dx}\left[\text{Si}(x^2)\right] = 2x\dfrac{\sin(x^2)}{x^2} = \dfrac{2\sin x^2}{x}$.

17. (a) The most obvious feature of the graph of $y = \sin(x^2)$ is its symmetry about the y-axis. This means the function $g(x) = \sin(x^2)$ is an even function, i.e. for all x, $g(x) = g(-x)$. Since $\sin(x^2)$ is even, its antiderivative F must be odd, that is $F(-x) = -F(-x)$. This can be seen since if $F(t) = \int_0^t \sin(x^2)\,dx$,

$$F(-t) = \int_0^{-t} \sin(x^2)\,dx = -\int_{-t}^0 \sin(x^2)\,dx = -\int_0^t \sin(x^2)\,dx,$$

since the area from $-t$ to 0 is the same as the area from 0 to t. Thus $F(t) = -F(-t)$ and F is odd.

The second obvious feature of the graph of $y = \sin(x^2)$ is that it oscillates between -1 and 1 with a "period" which goes to zero as $|x|$ increases. This implies that $F'(x)$ alternates between intervals where it is positive or negative, and increasing or decreasing, with frequency growing arbitrarily large as $|x|$ increases. Thus $F(x)$ itself similarly alternates between intervals where it is increasing or decreasing, and concave up or concave down.

Finally, since $y = \sin(x^2) = F'(x)$ passes through $(0,0)$, and $F(0) = 0$, F is tangent to the x-axis at the origin.

   (b)

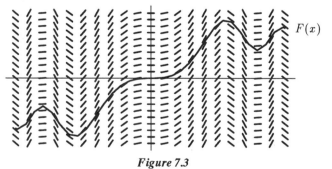

Figure 7.3

F never crosses the x-axis in the region $x > 0$, and $\displaystyle\lim_{x \to \infty} F(x)$ exists. One way to see these facts is to note that by the Construction Theorem,

$$F(x) = F(x) - F(0) = \int_0^x F'(t)\,dt.$$

So $F(x)$ is just the area under the curve $y = \sin(t^2)$ for $0 \le t \le x$. Now looking at the graph of curve, we see that this area will include alternating pieces above and below the x-axis. We can also see that the area of these pieces is approaching 0 as we go further out. So we add a piece, take a piece away, add another piece, take another piece away, and so on. It turns out that this means that the sums of the pieces converge. To see this, think of walking from point A to point B. If you walk almost to B, then go a smaller distance toward A, then a yet smaller distance back toward B, and so on, you will eventually approach some point between A and B. So we can see that $\lim_{x \to \infty} F(x)$ exists. Also, since we always subtract a smaller piece than we just added, and the first piece is added instead of subtracted, we see that we never get a negative sum; thus $F(x)$ is never negative in the region $x > 0$, so $F(x)$ never crosses the x-axis there.

21. Let $w^2 = \dfrac{t^2}{2}$ and $w = \dfrac{1}{\sqrt{2}}t$. Then $dt = \sqrt{2}\, dw$, and $w = \dfrac{x}{\sqrt{2}}$ when $t = x$.

Therefore $\sqrt{\dfrac{2}{\pi}} \displaystyle\int_0^x e^{-\frac{t^2}{2}}\, dt - \sqrt{\dfrac{2}{\pi}} \cdot \sqrt{2} \int_0^{x/\sqrt{2}} e^{-w^2}\, dw = \operatorname{erf}\left(\dfrac{x}{\sqrt{2}}\right)$.

SOLUTIONS TO REVIEW PROBLEMS FOR CHAPTER SEVEN

1. The limits of integration are 0 and b, and the rectangle represents the region under the curve $f(x) = h$ between these limits. Thus,

$$\text{Area of rectangle} = \int_0^b h\, dx = hx \Big|_0^b = hb.$$

5. (a) Recall that $x = e^{\ln x}$. Thus $x^x = (e^{\ln x})^x = e^{x \ln x}$.

   (b)
   $$\frac{d}{dx}(x^x) = \frac{d}{dx}(e^{x \ln x}) = e^{x \ln x} \frac{d}{dx}(x \ln x) \text{ by the chain rule}$$
   $$= e^{x \ln x}(\ln x + 1)$$
   $$= x^x(\ln x + 1).$$

   (c) By the Fundamental Theorem of Calculus and part (b),

   $$\int x^x(1 + \ln x)\, dx = x^x + C.$$

   (d) By the Fundamental Theorem of Calculus,

   $$\int_1^2 x^x(1 + \ln x)\, dx = x^x \Big|_1^2$$
   $$= 2^2 - 1^1$$
   $$= 3.$$

Using a calculator, we can check our answer numerically. With 50 subdivisions, the left-hand sum ≈ 2.943 and the right-hand sum ≈ 3.058. With 100 subdivisions, the left-hand sum ≈ 2.971 and the right-hand sum ≈ 3.029.

9. Let $w = 2 + 3\cos x$, so $dw = -3\sin x \, dx$, giving $-\frac{1}{3} dw = \sin x \, dx$. Then

$$\int \sin x \left(\sqrt{2+3\cos x}\right) dx = \int \sqrt{w} \left(-\frac{1}{3}\right) dw = -\frac{1}{3} \int \sqrt{w} \, dw$$

$$= \left(-\frac{1}{3}\right) \frac{w^{\frac{3}{2}}}{\frac{3}{2}} + C = -\frac{2}{9}(2 + 3\cos x)^{\frac{3}{2}} + C.$$

13. We integrate by parts, with $u = y$, $v' = \sin y$. We have $u' = 1$, $v = -\cos y$, and

$$\int y \sin y \, dy = -y \cos y - \int (-\cos y) \, dy = -y \cos y + \sin y + C.$$

Check:

$$\frac{d}{dy}(-y\cos y + \sin y + C) = -\cos y + y \sin y + \cos y = y \sin y.$$

17. Remember that $\ln(x^2) = 2\ln x$. Therefore,

$$\int \ln(x^2)\, dx = 2 \int \ln x \, dx = 2x \ln x - 2x + C.$$

Check:

$$\frac{d}{dx}(2x \ln x - 2x + C) = 2\ln x + \frac{2x}{x} - 2 = 2\ln x = \ln(x^2).$$

21. Substitute $w = 4 - x^2$, $dw = -2x\, dx$:

$$\int x\sqrt{4-x^2}\, dx = -\frac{1}{2}\int \sqrt{w}\, dw = -\frac{1}{3}w^{3/2} + C = -\frac{1}{3}(4-x^2)^{3/2} + C.$$

Check

$$\frac{d}{dx}\left[-\frac{1}{3}(4-x^2)^{3/2} + C\right] = -\frac{1}{3}\left[\frac{3}{2}(4-x^2)^{1/2}(-2x)\right] = x\sqrt{4-x^2}.$$

25.

$$\int \frac{(u+1)^3}{u^2} \, du = \int \frac{(u^3 + 3u^2 + 3u + 1)}{u^2} \, du$$

$$= \int \left(u + 3 + \frac{3}{u} + \frac{1}{u^2}\right) du$$

$$= \frac{u^2}{2} + 3u + 3\ln|u| - \frac{1}{u} + C.$$

Check:

$$\frac{d}{du}\left(\frac{u^2}{2} + 3u + 3\ln|u| - \frac{1}{u} + C\right) = u + 3 + 3/u + 1/u^2 = \frac{(u+1)^3}{u^2}.$$

29. $\int e^{2x}\sin 2x\, dx = \frac{1}{4}e^{2x}(\sin 2x - \cos 2x) + C$ by II- 8 in the integral table.

Thus $\int_{-\pi}^{\pi} e^{2x}\sin 2x = \left[\frac{1}{4}e^{2x}(\sin 2x - \cos 2x)\right]\Big|_{-\pi}^{\pi} = \frac{1}{4}(e^{-2\pi} - e^{2\pi}) \approx -133.8724.$

We get -133.37 using Simpson's rule with 10 intervals. With 100 intervals, we get -133.8724. Thus our answer matches the approximation of Simpson's rule.

33. After the substitution $w = 1 - x^2$, the first integral becomes

$$-\frac{1}{2}\int w^{-1}\, dw.$$

After the substitution $w = \ln x$, the second integral becomes

$$\int w^{-1}\, dw.$$

37. If $I(t)$ is average per capita income t years after 1987, then $I'(t) = r(t)$.

(a) Since $t = 8$ in 1995, by the Fundamental Theorem,

$$I(8) - I(0) = \int_0^8 r(t)\, dt = \int_0^8 480(1.024)^t\, dt$$

$$= \frac{480(1.024)^t}{\ln(1.024)}\Big|_0^8 = 4228$$

so $I(8) = 26{,}000 + 4228 = 30{,}228$.

(b)

$$I(t) - I(0) = \int_0^t r(t)\, dt = \int_0^t 480(1.024)^t\, dt$$

$$= \frac{480(1.024)^t}{\ln(1.024)}\Big|_0^t$$

$$= \frac{480}{\ln(1.024)}\left((1.024)^t - 1\right)$$

$$= 20{,}239\left((1.024)^t - 1\right)$$

Thus, since $I(0) = 26{,}000$,

$$I(t) = 26{,}000 + 20{,}239(1.024^t - 1)$$

$$= 20{,}239(1.024)^t + 5761$$

40. The point of intersection of the two curves $y = x^2$ and $y = 6 - x$ is at (2,4). The average height of the shaded area is the average value of the difference between the functions:

$$\frac{1}{(2 - 0)} \int_0^2 ((6 - x) - x^2)\, dx = \left(3x - \frac{x^2}{4} - \frac{x^3}{6} \right)\Big|_0^2$$

$$= \frac{11}{3}.$$

41. The average width of the shaded area is the average value of the horizontal distance between the two functions. If we call this horizontal distance $h(y)$, then the average width is

$$\frac{1}{(6 - 0)} \int_0^6 h(y)\, dy.$$

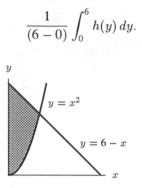

We could compute this integral if we wanted to, but we don't need to. We can simply note that the integral (without the $\frac{1}{6}$ term) is just the area of the shaded region; similarly, the integral in Problem 40 is *also* just the area of the shaded region. So they are the same. Now we know that our average width is just $\frac{1}{3}$ as much as the average height, since we divide by 6 instead of 2. So the answer is $\frac{11}{9}$.

45. $\int \frac{dx}{x \ln x} = \ln |\ln x| + C.$ (Substitute $w = \ln x$, $dw = \frac{1}{x}\, dx$).
Thus

$$\int_{10}^{\infty} \frac{dx}{x \ln x} = \lim_{b \to \infty} \int_{10}^b \frac{dx}{x \ln x} = \lim_{b \to \infty} \ln |\ln x| \Big|_{10}^b = \lim_{b \to \infty} \ln(\ln b) - \ln(\ln 10).$$

As $b \to \infty$, $\ln(\ln b) \to \infty$, so this diverges.

49. Since the value of $\tan \theta$ is between -1 and 1 on the interval $-\pi/4 \le \theta \le \pi/4$, our integral is not improper and so converges. Moreover, since $\tan \theta$ is an odd function, we have

$$\int_{-\frac{\pi}{4}}^{\frac{\pi}{4}} \tan \theta\, d\theta = \int_{-\frac{\pi}{4}}^0 \tan \theta\, d\theta + \int_0^{\frac{\pi}{4}} \tan \theta\, d\theta = -\int_{-\frac{\pi}{4}}^0 \tan(-\theta)\, d\theta + \int_0^{\frac{\pi}{4}} \tan \theta\, d\theta$$

$$= -\int_0^{\frac{\pi}{4}} \tan \theta\, d\theta + \int_0^{\frac{\pi}{4}} \tan \theta\, d\theta = 0.$$

53. This function is difficult to integrate, so instead we try to compare it with some other function. Since $\frac{\sin^2 \theta}{\theta^2+1} \geq 0$, we see that $\int_0^\infty \frac{\sin^2 \theta}{\theta^2+1} \, d\theta \geq 0$. Also, since $\sin^2 \theta \leq 1$,

$$\int_0^\infty \frac{\sin^2 \theta}{\theta^2 + 1} \, d\theta \leq \int_0^\infty \frac{1}{\theta^2 + 1} \, d\theta = \lim_{b \to \infty} \arctan \theta \Big|_0^b = \frac{\pi}{2}.$$

Thus $\int_0^\infty \frac{\sin^2 \theta}{\theta^2+1} \, d\theta$ converges, and its value is between 0 and $\frac{\pi}{2}$.

57. (a) $\displaystyle \int_0^\infty \sqrt{x}e^{-x} \, dx \approx 0.8862269 \ldots$ [It turns out that $\int_0^\infty \sqrt{x}e^{-x} \, dx = \frac{\sqrt{\pi}}{2}$]

(b) $\displaystyle \int_1^\infty \ln\left(\frac{e^x + 1}{e^x - 1}\right) \, dx = 0.747402 \ldots$

61. Let's assume that TRAP(10) and TRAP(50) are either both overestimates or both underestimates. Since TRAP(50) is more accurate, and it is bigger than TRAP(10), both are underestimates. Since TRAP(50) is 25 times more accurate, we have

$$I - \text{TRAP}(10) = 25(I - \text{TRAP}(50)),$$

where I is the value of the integral. Solving for I, we have

$$I \approx \frac{25 \, \text{TRAP}(50) - \text{TRAP}(10)}{24} \approx 4.6969$$

Thus the error for TRAP(10) is approximately 0.0078.

65.

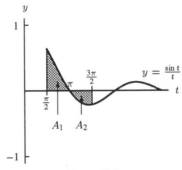

Figure 7.4

$F(x)$ represents the net area between $\frac{\sin t}{t}$ and the t-axis from $t = \frac{\pi}{2}$ to $t = x$, with area counted as negative for $\frac{\sin t}{t}$ below the t-axis. As long as the integrand is positive $F(x)$ is increasing. Therefore, the global maximum of $F(x)$ occurs at $x = \pi$ and is given by the area

$$A_1 = \int_{\pi/2}^\pi \frac{\sin t}{t} \, dt.$$

At $x = \pi/2$, $F(x) = 0$. Figure 7.4 shows that the area A_1 is larger than the area A_2. Thus $F(x) > 0$ for $\frac{\pi}{2} < x \leq \frac{3\pi}{2}$. Therefore the global minimum is $F(\frac{\pi}{2}) = 0$.

CHAPTER EIGHT

8.1 SOLUTIONS

1. (a) Suppose we choose an x, $0 \le x \le 2$. If Δx is a small fraction of a meter, then the density of the rod is approximately $\rho(x)$ anywhere from x to $x + \Delta x$ meters from the left end of the rod. The mass of the rod from x to $x + \Delta x$ meters is therefore approximately $\rho(x)\Delta x$.

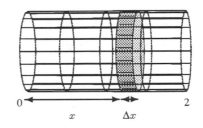

(b) The definite integral is

$$M = \int_0^2 \rho(x)\,dx = \int_0^2 (2 + 6x)\,dx = (2x + 3x^2)\Big|_0^2 = 16 \text{ grams}.$$

5. (a) Partition $0 \le r \le 8$ into eight subintervals of width $\Delta r = 1$ mile. Note that $r_i = i$. Let $y_i = \#$ of people living in the ith subinterval and y = the total population. Then $y_i \approx \rho(r_i)A_i$, where A_i = area in the ith subinterval. $A_i \approx 2\pi r_i \Delta r / 2 = \pi r_i = \pi i$. So $y_i \approx \rho(i) \cdot \pi i$, and the total population will be approximately

$$y \approx \sum_{i=0}^{7} \pi i \rho(i) = \pi[0(75) + 1(75) + 2(67.5) + 3(60) + 4(52.5) + 5(45) + 6(37.5) + 7(30)]$$

$$\approx 3.96 \text{ million people.}$$

(b) We expect our estimate to be an underestimate for several reasons. First, intuitively, the fact that the first term in our sum is 0, when we know the population in the first mile is some positive number, leads us to believe we are underestimating. Second, $(\pi r \rho(r))$, the function we are approximating, although not increasing over the whole interval $0 \le r \le 8$, is mostly increasing. We thus expect our left-hand sum to be an underestimate. Finally, A_i is actually $\pi(i + \frac{1}{2})$, not πi. (Check this.) Our underestimate in the areas (the A_i's) also causes our result to be an underestimate.

(c) For r between 1 and 8, $\rho(r) = 75 - 7.5(r - 1) = 82.5 - 7.5r$. Assuming the population density is constant for $0 \le r \le 1$, the number of people living within the first mile is $\frac{\pi}{2}(75)$ thousand. In the next seven miles, the total population is approximately

$$\int_1^8 \pi r(82.5 - 7.5r)\,dr = \left[\frac{82.5\pi}{2}r^2 - \frac{7.5\pi}{3}r^3\right]_1^8 \approx 4.15 \text{ million.}$$

Then $y \approx \left(\dfrac{\pi}{2} \dfrac{(75)}{1000} + 4.15 \right)$ million ≈ 4.27 million.

9. First we rewrite the chart, listing the density with the corresponding distance from the center of the earth (x km below the surface is equivalent to $6370 - x$ km from the center):

This gives us spherical shells whose volumes are $\frac{4}{3}\pi(r_i^3 - r_{i+1}^3)$ for any two consecutive distances from the origin. We will assume that the density of the earth is increasing with depth. Therefore, the average density of the ith shell is between D_i and D_{i+1}, the densities at top and bottom of shell i. So $\frac{4}{3}\pi D_{i+1}(r_i^3 - r_{i+1}^3)$ and $\frac{4}{3}\pi D_i(r_i^3 - r_{i+1}^3)$ are upper and lower bounds for the mass of the shell.

TABLE 8.1

i	x_i	$r_i = 6370 - x_i$	D_i
0	0	6370	3.3
1	1000	5370	4.5
2	2000	4370	5.1
3	2900	3470	5.6
4	3000	3370	10.1
5	4000	2370	11.4
6	5000	1370	12.6
7	6000	370	13.0
8	6370	0	13.0

To get a rough approximation of the mass of the earth, we don't need to use all the data. Let's just use the densities at $x = 0, 2900, 5000$ and 6370 km. Calculating an upper bound on the mass,

$$M_U = \frac{4}{3}\pi[13.0(1370^3 - 0^3) + 12.6(3470^3 - 1370^3) + 5.6(6370^3 - 3470^3)] \cdot 10^{15} \approx 7.29 \times 10^{27} \text{ g}.$$

The factor of 10^{15} may appear unusual. Remember the radius is given in kilometers and the density is given in g/cm^3, so we must convert kilometers to centimeters: 1 km $= 10^5$ cm , so 1 km$^3 = 10^{15}$ cm^3.

The lower bound is

$$M_L = \frac{4}{3}\pi[12.6(1370^3 - 0^3) + 5.6(3470^3 - 1370^3) + 3.3(6370^3 - 3470^3)] \cdot 10^{15} \approx 4.05 \times 10^{27} \text{ g}.$$

Here, our upper bound is just under 2 times our lower bound.

Using all our data, we can find a more accurate estimate. The upper and lower bounds are

$$M_U = \frac{4}{3}\pi \sum_{i=0}^{7} D_{i+1}(r_i^3 - r_{i+1}^3) \cdot 10^{15} \text{ g}$$

and

$$M_L = \frac{4}{3}\pi \sum_{i=0}^{7} D_i(r_i^3 - r_{i+1}^3) \cdot 10^{15} \text{ g}.$$

We have

$$M_U = \frac{4}{3}\pi[4.5(6370^3 - 5370^3) + 5.1(5370^3 - 4370^3) + 5.6(4370^3 - 3470^3)$$
$$+ 10.1(3470^3 - 3370^3) + 11.4(3370^3 - 2370^3) + 12.6(2370^3 - 1370^3)$$
$$+ 13.0(1370^3 - 370^3) + 13.0(370^3 - 0^3)] \cdot 10^{15} \text{ g}$$
$$\approx 6.50 \times 10^{27} \text{ g}.$$

and

$$M_L - \frac{4}{3}\pi[3.3(6370^3 - 5370^3) + 4.5(5370^3 - 4370^3) + 5.1(4370^3 - 3470^3)$$
$$+ 5.6(3470^3 - 3370^3) + 10.1(3370^3 - 2370^3) + 11.4(2370^3 - 1370^3)$$
$$+ 12.6(1370^3 - 370^3) + 13.0(370^3 - 0^3)] \cdot 10^{15} \text{ g}$$
$$\approx 5.46 \times 10^{27} \text{ g}.$$

8.2   SOLUTIONS

1. Vertical slices are circular. Horizontal slices would be similar to ellipses in cross-section, or at least ovals (a word derived from *ovum*, the Latin word for egg).

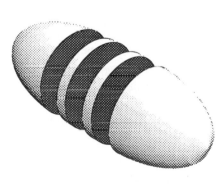
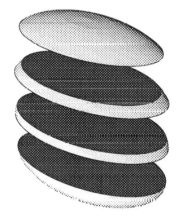

5.

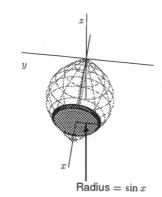

Radius $= \sin x$

We take slices perpendicular to the x–axis. The Riemann sum for approximating the volume is $\sum \pi \sin^2 x \Delta x$. The volume is the integral corresponding to that sum, namely

$$V = \int_0^\pi \pi \sin^2 x \, dx$$

$$= \pi \left[-\frac{1}{2} \sin x \cos x + \frac{1}{2} x \right] \Big|_0^\pi = \frac{\pi^2}{2} \approx 4.935.$$

9.

$r = e^x$

This is the volume of revolution gotten from the rotating the curve $y = e^x$. Take slices perpendicular to the x-axis. They will be circles with radius e^x, so

$$V = \int_{x=0}^{x=1} \pi y^2 \, dx = \pi \int_0^1 e^{2x} \, dx$$

$$= \frac{\pi e^{2x}}{2} \Big|_0^1 = \frac{\pi(e^2 - 1)}{2} \approx 10.036.$$

13.

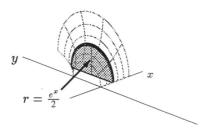

$r = \frac{e^x}{2}$

We slice perpendicular to the x-axis. As stated in the problem, the cross-sections obtained thereby will be semicircles, with radius $\frac{e^x}{2}$. The volume of one semicircular slice is $\frac{1}{2} \pi \left(\frac{e^x}{2} \right)^2 \, dx$. (Look at the picture.) Adding up the volumes of the slices yields

$$\text{Volume} = \int_{x=0}^{x=1} \pi \frac{y^2}{2} \, dx = \frac{\pi}{8} \int_0^1 e^{2x} \, dx$$

$$= \frac{\pi e^{2x}}{16} \Big|_0^1 = \frac{\pi(e^2 - 1)}{16} \approx 1.25.$$

17. The problem appears complicated, because we are now working in three dimensions. However, if we take one dimension at a time, we will see that the solution is not too difficult. For example, let's just work at a constant depth, say 0. We apply the trapezoid rule to find the approximate area along the length of the boat. For example, by the trapezoid rule the approximate area at depth 0 from the front of the boat to 10 feet toward the back is $\frac{(2+8)\cdot 10}{2} = 50$. Overall, at depth 0 we have that the area for each length span is as follows:

TABLE 8.2

length span:	0–10	10–20	20–30	30–40	40–50	50–60
depth 0	50	105	145	165	165	130

We can fill in the whole chart the same way:

TABLE 8.3

length span:		0–10	10–20	20–30	30–40	40–50	50–60
	0	50	105	145	165	165	130
	2	25	60	90	105	105	90
depth	4	15	35	50	65	65	50
	6	5	15	25	35	35	25
	8	0	5	10	10	10	10

Now, to find the volume, we just apply the trapezoid rule to the depths and areas. For example, according to the trapezoid rule the approximate volume as the depth goes from 0 to 2 and the length goes from 0 to 10 is $\frac{(50+25)\cdot 2}{2} = 75$. Again, we fill in a chart:

TABLE 8.4

length span:		0–10	10–20	20–30	30–40	40–50	50–60
	0–2	75	165	235	270	270	220
depth	2–4	40	95	140	170	170	140
span	4–6	20	50	75	100	100	75
	6–8	5	20	35	45	45	35

Adding all this up, we find the volume is approximately 2595 cubic feet.

You might wonder what would have happened if we had done our trapezoids along the depth axis first instead of along the length axis. If you try this, you'd find that you come up with the same answers in the volume chart! For the trapezoid rule, it doesn't matter which axis you choose first.

21.   Since $y = (e^x + e^{-x})/2$, $y' = (e^x - e^{-x})/2$. The length of the catenary is

$$\int_{-1}^{1} \sqrt{1 + (y')^2} \, dx = \int_{-1}^{1} \sqrt{1 + \left[\frac{e^x - e^{-x}}{2}\right]^2} \, dx = \int_{-1}^{1} \sqrt{1 + \frac{e^{2x}}{4} - \frac{1}{2} + \frac{e^{-2x}}{4}} \, dx$$

$$= \int_{-1}^{1} \sqrt{\left[\frac{e^x + e^{-x}}{2}\right]^2} \, dx = \int_{-1}^{1} \frac{e^x + e^{-x}}{2} \, dx$$

$$= \left[\frac{e^x - e^{-x}}{2}\right]\Bigg|_{-1}^{1} = e - e^{-1} \approx 2.35.$$

25.   (a)   If $f(x) = \int_0^x \sqrt{g'(t)^2 - 1} \, dt$, then, by the Fundamental Theorem of Calculus, $f'(x) = \sqrt{g'(x)^2 - 1}$. So the arc length of f from 0 to x is

$$\int_0^x \sqrt{1 + (f'(t))^2} \, dt = \int_0^x \sqrt{1 + (\sqrt{g'(t)^2 - 1})^2} \, dt$$

$$= \int_0^x \sqrt{1 + g'(t)^2 - 1} \, dt$$

$$= \int_0^x g'(t) \, dt = g(x) - g(0) = g(x).$$

(b)   If g is the arc length of any function f, then by the Fundamental Theorem of Calculus, $g'(x) = \sqrt{1 + f'(x)^2} \geq 1$. So if $g'(x) < 1$, g cannot be the arc length of a function.

(c)   We find a function f whose arc length from 0 to x is $g(x) = 2x$. Using part (a), we see that

$$f(x) = \int_0^x \sqrt{(g'(t))^2 - 1} \, dt = \int_0^x \sqrt{2^2 - 1} \, dt = \sqrt{3}x.$$

This is the equation of a line. Does it make sense to you that the arc length of a line segment depends linearly on its right endpoint?

8.3 SOLUTIONS

1. Let x be the distance measured from the bottom the tank. It follows that $0 \leq x \leq 10$. To pump a layer of water of thickness Δx at x feet from the bottom, the work needed is $62.4\pi 6^2(20 - x)\Delta x$. Therefore, the total work is

$$
\begin{aligned}
W &= \int_0^{10} 36 \cdot 62.4\pi(20 - x)dx \\
&= 36 \cdot 62.4\pi\left(20x - \frac{1}{2}x^2\right)\Big|_0^{10} \\
&= 36 \cdot 62.4\pi(200 - 50) \\
&\approx 1{,}058{,}591.1 \ \text{ft-lb.}
\end{aligned}
$$

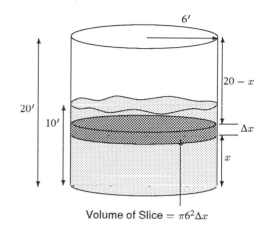

Volume of Slice $= \pi 6^2 \Delta x$

5. Let x be the height (in feet) from ground to the cube of ice. It follows that $0 \leq x \leq 100$. At height x, the ice cube w weighs $2000 - 4x$ since it's being lifted at a rate 1 ft./min. and it's melting at a rate of 4 lb/min. To lift it Δx more the work required is $(2000 - 4x)\Delta x$. So the total work done is

$$
\begin{aligned}
W &= \int_0^{100}(2000 - 4x)dx \\
&= (2000x - 2x^2)\Big|_0^{100} \\
&= 2000(100) - 2 \cdot (100)^2 \\
&= 180{,}000 \quad \text{ft-lb.}
\end{aligned}
$$

9. Setting the initial kinetic energy and escape work equal to each other gives

$$
\frac{1}{2}mv^2 = \frac{GMm}{R}, \text{ or } \quad v^2 = \frac{2GM}{R}.
$$

Since the planet is assumed to be a sphere of radius R and density ρ, we have $M = \rho(\frac{4}{3}\pi)R^3$. Hence

$$
v^2 = \frac{2G\rho(\frac{4}{3}\pi)R^3}{R}
$$

and therefore

$$
v = k\sqrt{\rho}R
$$

where $k = \sqrt{\frac{8\pi G}{3}}$. That is, the escape velocity is proportional to R and $\sqrt{\rho}$.

13.   We need to divide the disk up into circular rings of charge and integrate their contributions to the potential (at P) from 0 to a. These rings, however, are not uniformly distant from the point P. A ring of radius z is $\sqrt{R^2 + z^2}$ away from point P (See picture).

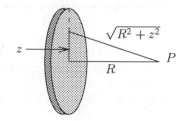

The ring has area $2\pi z \, \Delta z$, and charge $2\pi z \sigma \, \Delta z$. The potential of the ring is then $\dfrac{2\pi z \sigma \, \Delta z}{\sqrt{R^2 + z^2}}$ and the total potential at point P is

$$\int_0^a \frac{2\pi z \sigma \, dz}{\sqrt{R^2 + z^2}} = \pi \sigma \int_0^a \frac{2z \, dz}{\sqrt{R^2 + z^2}}.$$

We make the substitution $u = z^2$. Then $du = 2z \, dz$. We obtain

$$\pi \sigma \int_0^a \frac{2z \, dz}{\sqrt{R^2 + z^2}} = \pi \sigma \int_0^{a^2} \frac{du}{\sqrt{R^2 + u}} = \pi \sigma (2\sqrt{R^2 + u}) \Big|_0^{a^2}$$

$$= \pi \sigma (2\sqrt{R^2 + z^2}) \Big|_0^a = 2\pi \sigma (\sqrt{R^2 + a^2} - R).$$

(The substitution $u = R^2 + z^2$ or $\sqrt{R^2 + z^2}$ works also.)

17.

The density of the rod, in mass per unit length, is M/l. So a slice of size dr has mass $\frac{M \, dr}{l}$. It pulls the small mass m with force $Gm\frac{M \, dr}{l}/r^2 = \frac{GmM \, dr}{lr^2}$. So the total gravitational attraction between the rod and point is

$$\int_a^{a+l} \frac{GmM \, dr}{lr^2} = \frac{GmM}{l}\left(-\frac{1}{r}\right)\Big|_a^{a+l}$$

$$= \frac{GmM}{l}\left(\frac{1}{a} - \frac{1}{a+l}\right)$$

$$= \frac{GmM}{l}\frac{l}{a(a+l)} = \frac{GmM}{a(a+l)}.$$

8.4 SOLUTIONS

1.

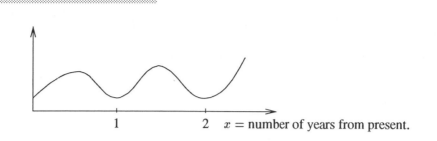

$x =$ number of years from present.

The graph reaches a peak each summer, and a trough each winter. The graph shows sunscreen sales increasing from cycle to cycle. This gradual increase may be due in part to inflation and to population growth.

5. (a) Solve for $P(t) = P$.

$$100000 = \int_0^{10} Pe^{0.10(10-t)}dt = Pe \int_0^{10} e^{-0.10t}dt$$

$$= \frac{Pe}{-0.10} e^{-0.10t} \Big|_0^{10} = Pe(-3.678 + 10)$$

$$= P \cdot 17.183$$

So, $P \approx \$5820$ per year.

(b) To answer this, we'll calculate the present value of \$100,000:

$$100000 = Pe^{0.10(10)}$$

$$P \approx \$36,787.94.$$

9. (a) Let's split the time interval into n parts, each of length Δt.

During the interval from t_i to t_{i+1}, profit is earned at a rate of approximately $(2 - 0.1t_i)$ thousand dollars per year, or $(2000 - 100t_i)$ dollars per year. Thus during this period, a total profit of $(2000 - 100t_i)\Delta t$ dollars is earned. Since this profit is earned t_i years in the future, its present value is $(2000 - 100t_i)\Delta t e^{-0.1t_i}$ dollars. Thus

$$\text{Total Present Value} \approx \sum_{i=0}^{n-1}(2000 - 100t_i)e^{-0.1t_i}\Delta t.$$

(b)   The Riemann sum corresponds to the integral:

$$\int_0^T e^{-0.10t}(2000 - 100t)\,dt.$$

(c)   To find where the present value is maximized, we take the derivative of

$$P(T) = \int_0^T e^{-0.10t}(2000 - 100t)\,dt,$$

and obtain

$$P'(T) = e^{-0.10T}(2000 - 100T).$$

This is 0 exactly when $2000 - 100T = 0$, that is, when $T = 20$ years. The value $T = 20$ maximizes $P(T)$, since $P'(T) > 0$ for $T < 20$, and $P'(T) < 0$ for $T > 20$. To determine what the maximum is, we evaluate the integral representation for $P(T)$ by formula III-14 in the integral table:

$$P(20) = \int_0^{20} e^{-0.10t}(2000 - 100t)\,dt$$

$$= \left[\frac{(2000 - 100t)}{-0.10}e^{-0.10t} + 10000e^{-0.10t}\right]\Big|_0^{20} \approx \$11353.35.$$

13.

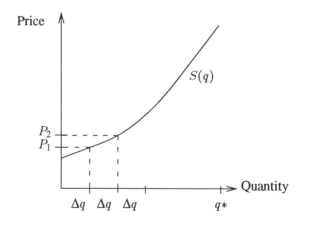

The supply curve, $S(q)$, represents the minimum price p per unit that the suppliers will be willing to supply some quantity q of the good for. If the suppliers have q^* of the good and q^* is divided into subintervals of size Δq, then if the consumers could offer the suppliers for each Δq a price increase just sufficient to induce the suppliers to sell an additional Δq of the good, the consumers' total expenditure on q^* goods would be

$$p_1\Delta q + p_2\Delta q + \cdots = \sum p_i\Delta q.$$

As $\Delta q \to 0$ the Riemann sum becomes the integral $\int_0^{q^*} S(q)\, dq$. Thus $\int_0^{q^*} S(q)\, dq$ is the amount the consumers would pay if suppliers could be forced to sell at the lowest price they would be willing to accept.

8.5   SOLUTIONS

1.

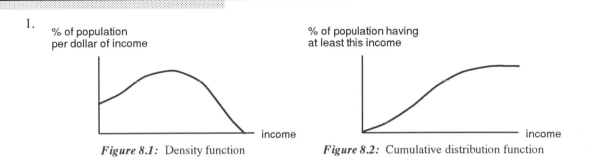

Figure 8.1: Density function   Figure 8.2: Cumulative distribution function

5. (a) Let $P(x)$ be the cumulative distribution function of the heights of the unfertilized plants. As do all cumulative distribution functions, $P(x)$ rises from 0 to 1 as x increases. The greatest number of plants will have heights in the range where $P(x)$ rises the most. The steepest rise appears to occur at about $x = 1$ m. Reading from the graph we see that $P(0.9) \approx 0.2$ and $P(1.1) \approx 0.8$, so that approximately $P(1.1) - P(0.9) = 0.8 - 0.2 = 0.6 = 60\%$ of the unfertilized plants grow to heights between 0.9 m and 1.1 m. Most of the plants grow to heights in the range 0.9 m to 1.1 m.

(b) Let $P_A(x)$ be the cumulative distribution function of the plants that were fertilized with A. Since $P_A(x)$ rises the most in the range 0.7 m $\leq x \leq$ 0.9 m, many of the plants fertilized with A will have heights in the range 0.7 m to 0.9 m. Reading from the graph of P_A, we find that $P_A(0.7) \approx 0.2$ and $P_A(0.9) \approx 0.8$, so $P_A(0.9) - P_A(0.7) \approx 0.8 - 0.2 = 0.6 = 60\%$ of the plants fertilized with A have heights between 0.7 m and 0.9 m. Fertilizer A had the effect of stunting the growth of the plants.

On the other hand, the cumulative distribution function $P_B(x)$ of the heights of the plants fertilized with B rises the most in the range 1.1 m $\leq x \leq$ 1.3 m, so most of these plants have heights in the range 1.1 m to 1.3 m. Fertilizer B caused the plants to grow about 0.2 m taller than they would have with no fertilizer.

9. (a) The fraction of students passing is given by the area under the curve from 2 to 4 divided by the total area under the curve. This appears to be about $\frac{2}{3}$.

(b) The fraction with honor grades corresponds to the area under the curve from 3 to 4 divided by the total area. This is about $\frac{1}{3}$.

(c) The peak around 2 probably exists because many students work to get just a passing grade.

(d)

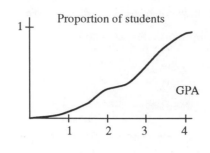

8.6 SOLUTIONS

1.

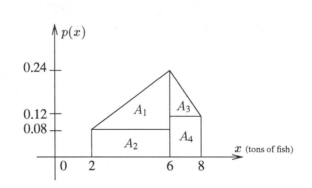

Splitting the figure into four pieces, we see that

$$\begin{aligned}
\text{Area under the curve} &= A_1 + A_2 + A_3 + A_4 \\
&= \frac{1}{2}(0.16)4 + 4(0.08) + \frac{1}{2}(0.12)2 + 2(0.12) \\
&= 1.
\end{aligned}$$

We expect the area to be 1, since $\int_{-\infty}^{\infty} p(x)\,dx = 1$ for any probability density function, and $p(x)$ is 0 except when $2 \leq x \leq 8$.

5. (a) i.   ii.

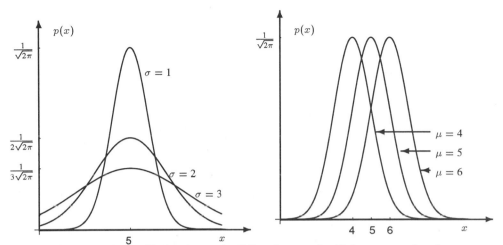

(b) Recall that the mean is the "balancing point." In other words, if the area under the curve was made of cardboard, we'd expect it to balance at the mean. All of the graphs are symmetric across the line $x = \mu$, so μ is the "balancing point" and hence the mean.

As the graphs also show, increasing σ flattens out the graph, in effect lessening the concentration of the data near the mean. Thus, the smaller the σ value, the more data is clustered around the mean.

9. (a) We want to find a such that $\int_0^\infty p(v)\,dv = \lim_{r \to \infty} a \int_0^r v^2 e^{-mv^2/2kT}\,dv = 1$. Therefore,

$$\frac{1}{a} = \lim_{r \to \infty} \int_0^r v^2 e^{-mv^2/2kT}\,dv. \tag{8.1}$$

To evaluate the integral, use integration by parts with the substitutions $u = v$ and $w' = ve^{-mv^2/2kT}$:

$$\int_0^r \underbrace{v}_{u}\,\underbrace{ve^{-mv^2/2kT}}_{w'}\,dv = \underbrace{v}_{u}\,\underbrace{\frac{e^{-mv^2/2kT}}{-m/kT}}_{w}\Bigg|_0^r - \int_0^r \underbrace{1}_{u'}\,\underbrace{\frac{e^{-mv^2/2kT}}{-m/kT}}_{w}\,dv$$

$$= -\frac{kTr}{m}e^{-mr^2/2kT} + \frac{kT}{m}\int_0^r e^{-mv^2/2kT}\,dv.$$

From the normal distribution we know that $\int_0^\infty \frac{1}{\sqrt{2\pi}}e^{-x^2/2}\,dx = \frac{1}{2}$, so

$$\int_0^\infty e^{-x^2/2}\,dx = \frac{\sqrt{2\pi}}{2}.$$

Therefore in the above integral, make the substitution $x = \sqrt{\frac{m}{kT}}\,v$, so that $dx = \sqrt{\frac{m}{kT}}\,dv$, or

$dv = \sqrt{\frac{kT}{m}}\, dx$. Then

$$\frac{kT}{m}\int_0^r e^{-mv^2/2kT}\, dv = \left(\frac{kT}{m}\right)^{3/2}\int_0^{\sqrt{\frac{m}{kT}}\, r} e^{-x^2/2}\, dx.$$

Substituting this into Equation 8.1 we get

$$\frac{1}{a} = \lim_{r\to\infty}\left(-\frac{kTr}{m}e^{-mr^2/2kT} + \left(\frac{kT}{m}\right)^{3/2}\int_0^{\sqrt{\frac{m}{kT}}\, r} e^{-x^2/2}\, dx\right) = 0 + \left(\frac{kT}{m}\right)^{3/2}\cdot\frac{\sqrt{2\pi}}{2}.$$

Therefore, $a = \frac{2}{\sqrt{2\pi}}\left(\frac{m}{kT}\right)^{3/2}$. Substituting the values for k, T, and m gives $a \approx 3.4 \times 10^{-8}$ SI units.

(b) To find the median, we wish to find the speed x such that

$$\int_0^x p(v)\, dv = \int_0^x av^2 e^{-\frac{mv^2}{2kT}}\, dv = \frac{1}{2},$$

where $a = \frac{2}{\sqrt{2\pi}}\left(\frac{m}{kT}\right)^{3/2}$. Using a calculator, by trial and error we get $x \approx 441$ m/sec.
To find the mean, we find

$$\int_0^\infty vp(v)\, dv = \int_0^\infty av^3 e^{-\frac{mv^2}{2kT}}\, dv.$$

This integral can be done by substitution. Let $u = v^2$, so $du = 2v\,dv$. Then

$$\int_0^\infty av^3 e^{-\frac{mv^2}{2kT}}\, dv = \frac{a}{2}\int_{v=0}^{v=\infty} v^2 e^{-\frac{mv^2}{2kT}}\, 2v\, dv$$

$$= \frac{a}{2}\int_{u=0}^{u=\infty} u e^{-\frac{mu}{2kT}}\, du$$

$$= \lim_{r\to\infty}\frac{a}{2}\int_0^r u e^{-\frac{mu}{2kT}}\, du.$$

Now, using the integral table, we have

$$\int_0^\infty av^3 e^{-\frac{mv^2}{2kT}}\, dv = \lim_{r\to\infty}\frac{a}{2}\left[-\frac{2kT}{m}u e^{-\frac{mu}{2kT}} - \left(-\frac{2kT}{m}\right)^2 e^{-\frac{mu}{2kT}}\right]\Bigg|_0^r$$

$$= \frac{a}{2}\left(-\frac{2kT}{m}\right)^2$$

$$\approx 457.7 \text{ m/sec.}$$

The maximum for $p(v)$ will be at a point where $p'(v) = 0$.

$$p'(v) = a(2v)e^{-\frac{mv^2}{2kT}} + av^2\left(-\frac{2mv}{2kT}\right)e^{-\frac{mv^2}{2kT}}$$

$$= ae^{-\frac{mv^2}{2kT}}\left(2v - v^3\frac{m}{kT}\right).$$

Thus $p'(v) = 0$ at $v = 0$ and at $v = \sqrt{\dfrac{2kT}{m}} \approx 405$. It's obvious that $p(0) = 0$, and that $p \to 0$ as $v \to \infty$. So $v = 405$ gives us a maximum: $p(405) \approx 0.002$.

(c) The mean, as we found in part (b), is $\dfrac{a}{2} \dfrac{4k^2T^2}{m^2} = \dfrac{4}{\sqrt{2\pi}} \dfrac{k^{1/2}T^{1/2}}{m^{1/2}}$. It is clear, then, that as T increases so does the mean. We found in part (b) that $p(v)$ reached its maximum at $v = \sqrt{\dfrac{2kT}{m}}$. Thus

$$\text{the maximum value of } p(v) = \frac{2}{\sqrt{2\pi}} \left(\frac{m}{kT}\right)^{3/2} \frac{2kT}{m} e^{-1}$$
$$= \frac{4}{e\sqrt{2\pi}} \frac{m^{1/2}}{kT^{1/2}}.$$

Thus as T increases, the maximum value decreases.

SOLUTIONS TO REVIEW PROBLEMS FOR CHAPTER EIGHT

1. (a)

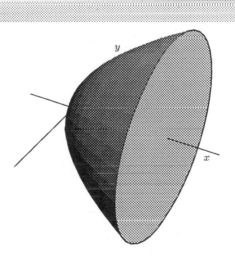

Figure 8.3: Rotated Region

(b) Divide [0,1] into N subintervals of width $\Delta x = \frac{1}{N}$. The volume of the ith disc is $\pi(\sqrt{x_i})^2 \Delta x = \pi x_i \Delta x$. So, $V \approx \sum_{i=1}^{N} \pi x_i \Delta x$.

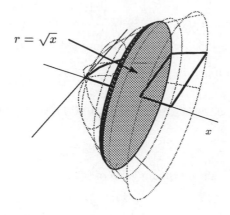

$r = \sqrt{x}$

Figure 8.4: Cutaway View

(c)

$$\text{Volume} = \int_0^1 \pi x \, dx = \left. \frac{\pi}{2} x^2 \right|_0^1 = \frac{\pi}{2} \approx 1.57 .$$

5. (a) Slice the headlight into N disks of height Δx by cutting perpendicular to the x–axis. The radius of each disk is y; the height is Δx. The volume of each disk is $\pi y^2 \Delta x$. Therefore, the Riemann sum approximating the volume of the headlight is

$$\sum_{i=1}^N \pi y_i^2 \Delta x = \sum_{i=1}^N \pi \frac{9 x_i}{4} \Delta x .$$

(b)

$$\pi \int_0^4 \frac{9x}{4} \, dx = \left. \pi \frac{9}{8} x^2 \right|_0^4 = 18\pi .$$

6.

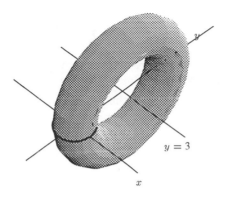

Figure 8.5: The Torus

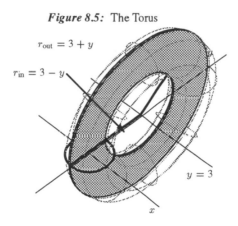

Figure 8.6: Slice of Torus

As shown in Figure 8.6, we slice the torus perpendicular to the line $y = 3$. We obtain washers with width dx, inner radius $r_{\text{in}} = 3 - y$, and outer radius $r_{\text{out}} = 3 + y$. Therefore, the area of the washer is $\pi r_{\text{out}}^2 - \pi r_{\text{in}}^2 = \pi[(3 + y)^2 - (3 - y)^2] = 12\pi y$. Since $y = \sqrt{1 - x^2}$, the volume is gotten by summing up the volumes of the washers: we get

$$\int_{-1}^{1} 12\pi\sqrt{1 - x^2}\, dx = 12\pi\int_{-1}^{1}\sqrt{1 - x^2}\, dx.$$

But $\int_{-1}^{1}\sqrt{1 - x^2}\, dx$ is the area of a semicircle of radius 1, which is $\frac{\pi}{2}$. So we get $12\pi\cdot\frac{\pi}{2} = 6\pi^2 \approx 59.22$. (Or, you could use

$$\int\sqrt{1 - x^2}\, dx = \left[x\sqrt{1 - x^2} + \arcsin(x)\right],$$

by Formula VI- 31 and Formula VI- 28.)

9.

$$L = \int_1^2 \sqrt{1 + e^{2x}}\, dx \approx 4.79.$$

Note that $\sqrt{1 + e^{2x}}$ does not have an obvious elementary antiderivative, so we use an approximation method to find an approximate value for L.

13. We'll find the arc length of the top half of the ellipse, and multiply that by 2. In the top half of the ellipse, the equation $(x^2/a^2) + (y^2/b^2) = 1$ implies

$$y = +b\sqrt{1 - \frac{x^2}{a^2}}.$$

Differentiating $(x^2/a^2) + (y^2/b^2) = 1$ implicitly with respect to x gives us

$$\frac{2x}{a^2} + \frac{2y}{b^2}\frac{dy}{dx} = 0,$$

so

$$\frac{dy}{dx} = \frac{\frac{-2x}{a^2}}{\frac{2y}{b^2}} = -\frac{b^2 x}{a^2 y}.$$

Substituting this into the arc length formula, we get

$$\text{Arc Length} = \int_{-a}^{a} \sqrt{1 + \left(-\frac{b^2 x}{a^2 y}\right)^2}\, dx$$

$$= \int_{-a}^{a} \sqrt{1 + \left(\frac{b^4 x^2}{a^4 (b^2)(1 - \frac{x^2}{a^2})}\right)}\, dx$$

$$= \int_{-a}^{a} \sqrt{1 + \left(\frac{b^2 x^2}{a^2 (a^2 - x^2)}\right)}\, dx.$$

Hence the arc length of the entire ellipse is

$$2\int_{-a}^{a} \sqrt{1 + \left(\frac{b^2 x^2}{a^2 (a^2 - x^2)}\right)}\, dx.$$

17. (a) The area under the graph of the height density function $p(x)$ is concentrated in two humps centered at 0.5 m and 1.1 m. The plants can therefore be separated into two groups, those with heights in the range 0.3 m to 0.7 m, corresponding to the first hump, and those with heights in the range 0.9 m to 1.3 m, corresponding to the second hump. This grouping of the grasses according to height is probably close to the species grouping. Since the second hump contains more area than the first, there are more plants of the tall grass species in the meadow.

(b) As do all cumulative distribution functions, the cumulative distribution function $P(x)$ of grass heights rises from 0 to 1 as x increases. Most of this rise is achieved in two spurts, the first as x goes from 0.3 m to 0.7 m, and the second as x goes from 0.9 m to 1.3 m. The plants can therefore be separated into two groups, those with heights in the range 0.3 m to 0.7 m, corresponding to the first spurt, and those with heights in the range 0.9 m to 1.3 m, corresponding to the second spurt. This grouping of the grasses according to height is the same as the grouping we made in part (a), and is probably close to the species grouping.

(c) The fraction of grasses with height less than 0.7 m equals $P(0.7) = 0.25 = 25\%$. The remaining 75% are the tall grasses.

21. Let x be the height from ground to the weight. It follows that $0 \le x \le 20$. At height x, to lift the weight Δx more, the work needed is $200\Delta x + 2(20 - x)\Delta x = (240 - 2x)\Delta x$. So the total work is

$$W = \int_0^{20} (240 - 2x)\,dx$$

$$= (240x - x^2)\Big|_0^{20}$$

$$= 240(20) - 20^2 = 4400 \quad \text{ft-lb.}$$

25. Bottom: The pressure on the bottom of the can is $2 \cdot 62.4 = 124.8$ pounds per square foot. The area of the bottom is $\pi r^2 = \pi$. So the force on the bottom of the can is $197.2 \times \pi \approx 618.3$ pounds. (This is just the weight of all the water.)

Side: To find the total force on the side of the can, we unfold the side to form a rectangle whose length is the circumference of the base, 2π feet. (This unfolding is like peeling off a label.) Next, we divide the label into strips of height dy. The area of each strip is $2\pi\,dy$ square feet, and the pressure at depth y is $62.4y$ pounds per square foot.

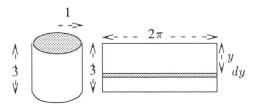

The force on each strip, therefore, is

$$(62.4y)(2\pi\,dy) \approx 392.07y\,dy \text{ pounds.}$$

So the force on the whole side is approximately

$$\int_0^2 392.07y\,dy = 392.07\frac{y^2}{2}\Big|_0^2 = 794.14 \text{ pounds.}$$

29. Any small piece of mass ΔM on either of the two spheres has kinetic energy $\frac{1}{2}v^2\Delta M$. Since the angular velocity of the two spheres is the same, the actual velocity of the piece ΔM will depend on how far away it is from the axis of revolution. The further away a piece is from the axis, the faster it must be moving and the larger its velocity v. This is because if ΔM is at a distance r from the axis, in one revolution it must trace out a circular path of length $2\pi r$ about the axis. Since every piece in either sphere takes 1 minute to make 1 revolution, pieces farther from the axis must move faster, as they travel a greater distance.

Thus, since the thin spherical shell has more of its mass concentrated farther from the axis of rotation than does the solid sphere, the bulk of it is traveling faster than the bulk of the solid sphere. So, it has the higher kinetic energy.

33.

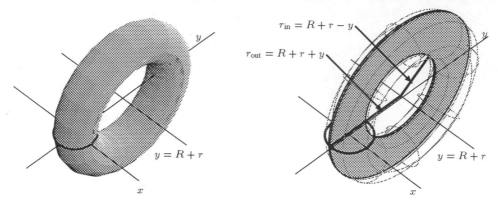

Figure 8.7: The Bagel *Figure 8.8:* Slice of Bagel

Looking back on Problem 6, we see one way to approach the problem. We will get a bagel–also known as a torus–by rotating a circle of radius R centered at the origin around the line $y = r + R$. (See Fig 8.7 and Fig 8.8.) Notice that r is the inner radius of the torus, so $r = d/2$. Further, $r + 2R$ is the outer radius of the torus, so $r + 2R = D/2$, or $R = \frac{1}{4}(D - d)$. Now, to find the volume of the torus, we slice it into rings of thickness Δx, perpendicular to the x-axis, just as we did in Problem 6.
The volume enclosed by the outer ring is $\pi(R + r + y)^2\,\Delta x$, and the volume enclosed by the inner ring is $\pi(R + r - y)^2\,\Delta x$, so the volume of a slice is

$$\pi(R + r + y)^2\,\Delta x - \pi(R + r - y)^2\,\Delta x = 4\pi(R + r)y\,\Delta x.$$

On the original circle, centered at the origin, $y = \sqrt{R^2 - x^2}$, so the total volume can be found by integrating:

$$V = \int_{-R}^{R} 4\pi(R + r)\sqrt{R^2 - x^2}\,dx$$

$$= 4\pi(R + r)\left[\frac{1}{2}x\sqrt{R^2 - x^2} + R^2\arcsin\frac{x}{R}\right]\Bigg|_{-R}^{R}$$

$$= 2\pi(R + r)R^2 \left(\frac{\pi}{2} - \left(-\frac{\pi}{2}\right)\right) = 2\pi^2(R + r)R^2,$$

using Formula 30 of the integral table to find $\int \sqrt{R^2 - x^2}\, dx$. Substituting for R and r, we get

$$V = 2\pi^2 \left[\frac{1}{4}(D + d)\right] \left[\frac{1}{4}(D - d)\right]^2 = \frac{1}{32}\pi^2(D^2 - d^2)(D - d).$$

CHAPTER NINE

9.1 SOLUTIONS

1. (a) = (III), (b) = (IV), (c) = (I), (d) = (II).

5. If $y = \cos \omega t$, then

$$\frac{dy}{dt} = -\omega \sin \omega t, \quad \frac{d^2 y}{dt^2} = -\omega^2 \cos \omega t.$$

Thus, if $\frac{d^2 y}{dt^2} + 9y = 0$, then

$$-\omega^2 \cos \omega t + 9 \cos \omega t = 0$$
$$(9 - \omega^2) \cos \omega t = 0.$$

Thus $9 - \omega^2 = 0$, or $\omega^2 = 9$, so $\omega = \pm 3$.

9.

$$(I) \ y = 2 \sin x, \qquad dy/dx = 2 \cos x, \qquad d^2 y/dx^2 = -2 \sin x$$
$$(II) \ y = \sin 2x, \qquad dy/dx = 2 \cos 2x, \qquad d^2 y/dx^2 = -4 \sin 2x$$
$$(III) \ y = e^{2x}, \qquad dy/dx = 2e^{2x}, \qquad d^2 y/dx^2 = 4e^{2x}$$
$$(IV) \ y = e^{-2x}, \qquad dy/dx = -2e^{-2x}, \qquad d^2 y/dx^2 = 4e^{-2x}$$

and so:

(a) (IV)
(b) (III)
(c) (III), (IV)
(d) (II)

9.2 SOLUTIONS

1. (a)

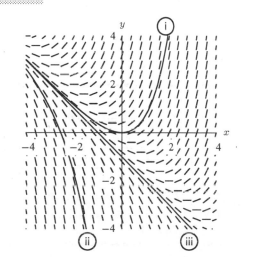

(b) The solution through $(-1, 0)$ appears to be linear with the equation $y = -x - 1$.

(c) If $y = -x - 1$, then $y' = -1$ and $x + y = x + (-x - 1) = -1$.

5. The first graph has the equation $y' = x^2 - y^2$. We can see this by looking along the line $y = x$. On the first slope field, it seems that $y' = 0$ along this line, as it should if $y' = x^2 - y^2$. This is not the case for the second graph.

At $(0, 1)$, $y' = -1$, and at $(1, 0)$, $y' = 1$, so you are looking for points on the axes where the line is sloped at $45°$.

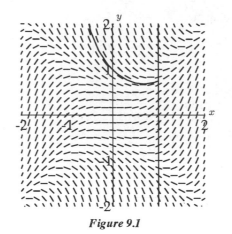

Figure 9.1

9. If the starting point has $y > 0$, then $y \to \infty$ as $x \to \infty$. If the starting point has $y = 0$, then y stays at 0 as $x \to \infty$. If the starting point has $y < 0$, then $y \to -\infty$ as $x \to \infty$.

13. If $y = 4$ for the starting point, then $y = 4$ always, so $y = 4$ as $x \to \infty$. If $y \neq 4$ for the starting point, then $y \to 4$ as $x \to \infty$.

9.3 SOLUTIONS

1. (a) (i)

TABLE 9.1 *Euler's method for*
$y' = (\sin x)(\sin y)$, *starting at* $(0, 2)$

x	y	$\Delta y =$ (slope)Δx
0	2	$0 = (\sin 0)(\sin 2)(0.1)$
0.1	2	$0.009 = (\sin 0.1)(\sin 2)(0.1)$
0.2	2.009	$0.018 = (\sin 0.2)(\sin 2.009)(0.1)$
0.3	2.027	

(ii)

TABLE 9.2 *Euler's method for*
$y' = (\sin x)(\sin y)$, *starting at* $(0, \pi)$

x	y	$\Delta y =$(slope)Δx
0	π	$0 = (\sin 0)(\sin \pi)(0.1)$
0.1	π	$0 = (\sin 0.1)(\sin \pi)(0.1)$
0.2	π	$0 = (\sin 0.2)(\sin \pi)(0.1)$
0.3	π	

(b) The slope field shows that the slope of the solution curve through $(0, \pi)$ is always 0. Thus the solution curve is the horizontal line with equation $y = \pi$.

5. (a) $\Delta x = 0.5$

TABLE 9.3 *Euler's method for*
$y' = 2x$, *with* $y(0) = 1$

x	y	$\Delta y =$(slope)Δx
0	1	$0 = (2 \cdot 0)(0.5)$
0.5	1	$0.5 = (2 \cdot 0.5)(0.5)$
1	1.5	

(b) $\Delta x = 0.25$

TABLE 9.4 *Euler's method for* $y' = 2x$,
with $y(0) = 1$

x	y	$\Delta y =$(slope)Δx
0	1	$0 = (2 \cdot 0)(0.25)$
0.25	1	$0.125 = (2 \cdot 0.25)(0.25)$
0.50	1.125	$0.25 = (2 \cdot 0.5)(0.25)$
0.75	1.375	$0.375 = (2 \cdot 0.75)(0.25)$
1	1.75	

(c) General Solution is $y = x^2 + C$, and $y(0) = 1$ gives $C = 1$. Thus, the solution is $y = x^2 + 1$.

(d) True value of y when $x = 1$ is $y = 1^2 + 1 = 2$.
   When $\Delta x = 0.5$, error $= 0.5$.
   When $\Delta x = 0.25$, error $= 0.25$.
   Thus, decreasing Δx by a factor of 2 has decreased the error by a factor of 2, as expected.

9. (a) Using one step, $\frac{\Delta B}{\Delta t} = 0.05$, so $\Delta B = \left(\frac{\Delta B}{\Delta t}\right) \Delta t = 50$. Therefore we get an approximation of $B \approx 1050$ after one year.

(b) With two steps, $\Delta t = 0.5$ and we have

TABLE 9.5

t	B	$\Delta B = (0.05 B)\Delta t$
0	1000	25
0.5	1025	25.63
1.0	1050.63	

(c) Keeping track to the nearest hundredth with $\Delta t = 0.25$, we have

TABLE 9.6

t	B	$\Delta B = (0.05 B)\Delta t$
0	1000	12.5
0.25	1012.5	12.66
0.5	1025.16	12.81
0.75	1037.97	12.97
1	1050.94	

(d) In part (a), we get our approximation by making a single increment, ΔB, where ΔB is just $0.05\,B$. If we think in terms of interest, ΔB is just like getting one end of the year interest payment. Since ΔB is 0.05 times the balance B, it is like getting 5% interest at the end of the year.

(e) Part (b) is equivalent to computing the final amount in an account that begins with $1000 and earns 5% interest compounded twice annually. Each step is like computing the interest after 6 months. When $t = 0.5$, for example, the interest is $\Delta B = (0.05B) \cdot \frac{1}{2}$, and we add this to $1000 to get the new balance.

Similarly, part (c) is equivalent to the final amount in an account that has an initial balance of $1000 and earns 5% interest compounded quarterly.

9.4 SOLUTIONS

1. $\frac{dP}{dt} = 0.02P$ implies that $\frac{dP}{P} = 0.02\,dt$.

$\int \frac{dP}{P} = \int 0.02\,dt$ implies that $\ln |P| = 0.02t + C$.

$|P| = e^{0.02t+C}$ implies that $P = Ae^{0.02t}$, where $A = \pm e^C$.
We are given $P(0) = 20$. Therefore, $P(0) = Ae^{(0.02)\cdot 0} = A = 20$. So the solution is $P = 20e^{0.02t}$.

5. $\frac{dy}{dx} + \frac{y}{3} = 0$ implies $\frac{dy}{dx} = -\frac{y}{3}$ implies $\int \frac{dy}{y} = -\int \frac{1}{3}\,dx$.
Integrating and moving terms, we have $y = Ae^{-\frac{1}{3}x}$. Since $y(0) = A = 10$, we have $y = 10e^{-\frac{1}{3}x}$.

9. Factoring out the 0.1 gives $\frac{dm}{dt} = 0.1m + 200 = 0.1(m + 2000)$.

   $\frac{dm}{m+2000} = 0.1\,dt$ implies that $\int \frac{dm}{m+2000} = \int 0.1\,dt$, so $\ln|m+2000| = 0.1t + C$. So $m = Ae^{0.1t} - 2000$. Using the initial condition, $m(0) = Ae^{(0.1)\cdot 0} - 2000 = 1000$, gives $A = 3000$. Thus $m = 3000e^{0.1t} - 2000$.

13. $\frac{dy}{dt} = y^2(1+t)$ implies that $\int \frac{dy}{y^2} = \int(1+t)\,dt$ implies that $-\frac{1}{y} = t + \frac{t^2}{2} + C$ implies that $y = -\frac{1}{t + \frac{t^2}{2} + C}$.

    Since $y = 2$ when $t = 1$, then $2 = -\frac{1}{1 + \frac{1}{2} + C}$. So $2C + 3 = -1$, and $C = -2$. Thus $y = -\frac{1}{\frac{t^2}{2} + t - 2} = -\frac{2}{t^2 + 2t - 4}$.

17. (a) Separating variables

$$\int \frac{dy}{100 - y} = \int dt$$
$$-\ln|100 - y| = t + C$$
$$|100 - y| = e^{-t-C}$$
$$100 - y = (\pm e^{-C})e^{-t} = Ae^{-t} \quad \text{where } A = \pm e^{-C}$$
$$y = 100 - Ae^{-t}.$$

(b)

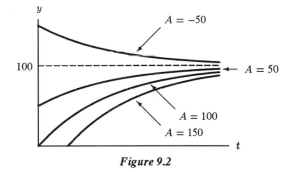

Figure 9.2

(c) Substituting $y = 0$ when $t = 0$ gives

$$0 = 100 - Ae^{-0}$$

so $A = 100$. Thus solution is

$$y = 100 - 100e^{-t}.$$

21. $\frac{dQ}{dt} = b - Q$ implies that $\frac{dQ}{b-Q} = dt$ which, in turn, implies $\int \frac{dQ}{b-Q} = \int dt$. Integrating yields $-\ln|b - Q| = t + C$, so $|b - Q| = e^{-(t+C)} = e^{-t}e^{-C}$. $Q = b - Ae^{-t}$, where $A = \pm e^{-C}$ or $A = 0$.

25. $t\frac{dx}{dt} = (1 + 2\ln t)\tan x$ implies that $\frac{dx}{\tan x} = (\frac{1+2\ln t}{t})\,dt$ which implies that $\int \frac{\cos x}{\sin x}\,dx = \int(\frac{1}{t} + \frac{2\ln t}{t})\,dt$. $\ln|\sin x| = \ln t + (\ln t)^2 + C$. (We can write $\ln t$, since $t > 0$.)

    $|\sin x| = e^{\ln t + (\ln t)^2 + C} = t(e^{\ln t})^{\ln t}e^C = t(t^{\ln t})e^C$. So $\sin x = At^{(\ln t + 1)}$, where $A = \pm e^C$. Therefore $x = \arcsin(At^{\ln t + 1})$.

29. (a) See (b).
     (b)

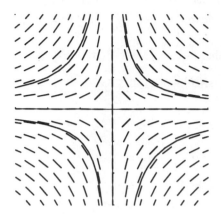

(c) $\frac{dy}{dx} = -\frac{y}{x}$, which implies that $\int \frac{dy}{y} = -\int \frac{dx}{x}$, so $\ln|y| = -\ln|x| + C$ implies that $|y| = e^{-\ln|x|+C} = (|x|)^{-1}e^C$.

$y = \frac{A}{x}$, where $A = \pm e^C$.

9.5 SOLUTIONS

1. (a) If the world's population grows exponentially, satisfying $\frac{dP}{dt} = kP$, and if the arable land used is proportional to the population, then we'd expect A to satisfy $\frac{dA}{dt} = kA$. One is, of course, also assuming that the amount of arable land is large compared to the amount that is now being used.
    (b) We must solve $A = A_0 e^{kt} = (1 \times 10^9)e^{kt}$, where t is the number of years after 1950. Since $2 \times 10^9 = (1 \times 10^9)e^{k(30)}$, we have $e^{30k} = 2$, so $k = \frac{\ln 2}{30} \approx 0.023$. Thus, $A \approx (1 \times 10^9)e^{0.023t}$. We want to find t such that $3.2 \times 10^9 = A = (1 \times 10^9)e^{0.023t}$. Taking logarithms yields

$$t = \frac{\ln(3.2)}{0.023} \approx 50.6 \text{ years.}$$

Thus the arable land will have run out by the year 2001.

5. (a) The rate of growth of the money in the account is proportional to the amount of money in the account. Thus

$$\frac{dM}{dt} = rM.$$

    (b) Solving, we have $\frac{dM}{M} = r\,dt$.

$$\int \frac{dM}{M} = \int r\,dt$$
$$\ln|M| = rt + C$$
$$M = e^{rt+C} = Ae^{rt}$$

When $t = 0$ (in 1970), $M = 1000$, so $A = 1000$ and $M = 1000e^{rt}$.

(c)

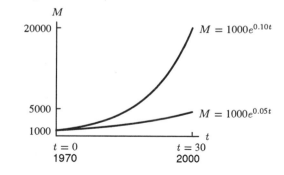

9. Lake Superior will take the longest, because the lake is largest (V is largest) and water is moving through it most slowly (r is smallest). Lake Erie looks as though it will take the least time because V is smallest and r is close to the largest. For Erie, $k = r/V = 175/460 = 0.38$. The lake with the largest value of r is Ontario, where $k = r/V = 209/1600 = 0.13$. Since e^{-kt} decreases faster for larger k, Lake Erie will take the shortest time for any fixed fraction of the pollution to be removed.

For Lake Superior

$$\frac{dQ}{dt} = -\frac{r}{V}Q = -\frac{65.2}{12,200}Q = -0.0053Q$$

so

$$Q = Q_0 e^{-0.0053t}.$$

When 80% of the pollution has been removed, 20% remains so $Q = 0.2Q_0$. Substituting

$$0.2Q_0 = Q_0 e^{-0.0053t}$$

so

$$t = -\frac{\ln(0.2)}{0.0053} \approx 301 \text{ years.}$$

(Note: The 301 is obtained by using the exact value of $\frac{r}{V} = \frac{65.2}{12200}$, rather than 0.0053. Using 0.0053 gives 304 years.) For Lake Erie, as in the text

$$\frac{dQ}{dt} = -\frac{r}{V}Q = -\frac{175}{460}Q = -0.38Q$$

so

$$Q = Q_0 e^{-0.38t}.$$

When 80% of the pollution has been removed

$$0.2Q_0 = Q_0 e^{-0.38t}$$

$$t = -\frac{\ln(0.2)}{0.38} \approx 4 \text{ years.}$$

So the ratio is

$$\frac{\text{Time for Lake Superior}}{\text{Time for Lake Erie}} \approx \frac{301}{4} \approx 75.$$

In other words it will take about 75 times as long to clean Lake Superior as Lake Erie.

13.   (a)   If I is intensity and l is the distance traveled through the water, then for some $k > 0$,

$$\frac{dI}{dl} = -kI.$$

(The proportionality constant is negative because intensity decreases with distance) Thus $I = Ae^{-kl}$. Since $I = A$ when $l = 0$, A represents the initial intensity of the light.

(b)   If 50% of the light is absorbed in 10 feet, then $0.50A = Ae^{-10k}$, so $e^{-10k} = \frac{1}{2}$, giving

$$k = \frac{-\ln\frac{1}{2}}{10} = \frac{\ln 2}{10}.$$

In 20 feet, the percentage of light left is

$$e^{-\frac{\ln 2}{10} \cdot 20} = e^{-2\ln 2} = \left(e^{\ln 2}\right)^{-2} = 2^{-2} = \frac{1}{4},$$

so $\frac{3}{4}$ or 75% of the light has been absorbed. Similarly, after 25 feet,

$$e^{-\frac{\ln 2}{10} \cdot 25} = e^{-2.5\ln 2} = \left(e^{\ln 2}\right)^{-\frac{5}{2}} = 2^{-\frac{5}{2}} \approx 0.177.$$

Approximately 17.7% of the light is left, so 82.3% of the light has been absorbed.

17.

$$\left(\begin{array}{c}\text{Rate at which quantity of}\\ \text{carbon-14 is increasing}\end{array}\right) = -k(\text{current quantity}).$$

If Q is the quantity of carbon-14 at time t (in years)

$$\text{Rate at which quantity is increasing} = \frac{dQ}{dt} = -kQ.$$

This differential equation has solution
$$Q = Q_0 e^{-kt}$$

where Q_0 is the initial quantity. Since at the end of one year 9999 parts are left out of 10,000, we know that
$$9999 = 10{,}000e^{-k(1)}.$$

Solving for k gives
$$k = \ln 0.9999 = 0.0001.$$

Thus $Q = Q_0 e^{-0.0001t}$. See Figure 9.3.

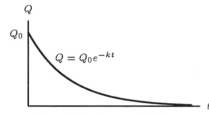

Figure 9.3: Exponential decay

9.6 SOLUTIONS

1. Let $D(t)$ be the quantity of dead leaves (in grams) per square centimeter. Then $\frac{dD}{dt} = 3 - 0.75D$, where t is in years. We factor out -0.75 and then separate variables.

$$\frac{dD}{dt} = -0.75(D - 4)$$

$$\int \frac{dD}{D - 4} = \int -0.75\, dt$$

$$\ln |D - 4| = -0.75t + C$$

$$|D - 4| = e^{-0.75t + C} = e^{-0.75t} e^C$$

$$D = 4 + Ae^{-0.75t}, \text{ where } A = \pm e^C.$$

If initially the ground is clear, the solution looks like:

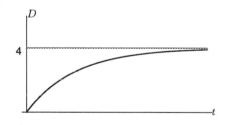

The equilibrium level is 4 grams per square centimeter, regardless of the initial condition.

5. Let the depth of the water at time t be y. Then $\frac{dy}{dt} = -k\sqrt{y}$, where k is a positive constant. Separating variables,

$$\int \frac{dy}{\sqrt{y}} = -\int k\, dt,$$

so

$$2\sqrt{y} = -kt + C.$$

When $t = 0$, $y = 36$; $2\sqrt{36} = -k \cdot 0 + C$, so $C = 12$.
When $t = 1$, $y = 35$; $2\sqrt{35} = -k + 12$, so $k \approx 0.17$.
Thus, $2\sqrt{y} \approx -0.17t + 12$. We are looking for t such that $y = 0$; this happens when $t \approx \frac{12}{0.17} \approx 71$
hours, or about 3 days.

9.   Using (rate balance increasing) = (rate interest added)− (rate payments made), when the interest rate
is i, we have

$$\frac{dB}{dt} = iB - 100.$$

Solving this equation, we find:

$$\frac{dB}{dt} = i\left(B - \frac{100}{i}\right)$$

$$\int \frac{dB}{B - \frac{100}{i}} = \int i\, dt$$

$$\ln\left|B - \frac{100}{i}\right| = it + C$$

$$B - \frac{100}{i} = Ae^{it}, \text{ where } A = \pm e^{C}.$$

At time $t = 0$ we start with a balance of \$1000. Thus
$1000 - \frac{100}{i} = Ae^{0}$, so $A = 1000 - \frac{100}{i}$.
Thus $B = \frac{100}{i} + (1000 - \frac{100}{i})e^{it}$.
When $i = 0.05$, $B = 2000 - 1000e^{0.05t}$.
When $i = 0.1$, $B = 1000$.
When $i = 0.15$, $B = 666.67 + 333.33e^{0.15t}$.
We now look at the graph when $i = 0.05$, $i = 0.1$, and $i = 0.15$.

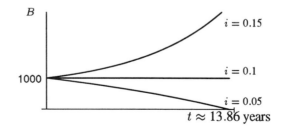

13.   (a)

$$\frac{dQ}{dt} = r - \alpha Q = -\alpha\left(Q - \frac{r}{\alpha}\right)$$

$$\int \frac{dQ}{Q - r/\alpha} = = -\alpha \int dt$$

$$\ln \left| Q - \frac{r}{\alpha} \right| = -\alpha t + C$$

$$Q - \frac{r}{\alpha} = Ae^{-\alpha t}$$

When $t = 0$, $Q = 0$, so $A = -\frac{r}{\alpha}$ and

$$Q = \frac{r}{\alpha}(1 - e^{-\alpha t})$$

So,

$$Q_\infty = \lim_{t \to \infty} Q = \frac{r}{\alpha}.$$

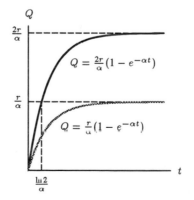

(b)   Doubling r doubles Q_∞. Since $Q_\infty = r/\alpha$, the time to reach $\frac{1}{2}Q_\infty$ is obtained by solving

$$\frac{r}{2\alpha} = \frac{r}{\alpha}(1 - e^{-\alpha t})$$

$$\frac{1}{2} = 1 - e^{-\alpha t}$$

$$e^{-\alpha t} = \frac{1}{2}$$

$$t = -\frac{\ln(1/2)}{\alpha} = \frac{\ln 2}{\alpha}.$$

So altering r doesn't alter the time it takes to reach $\frac{1}{2}Q_\infty$.

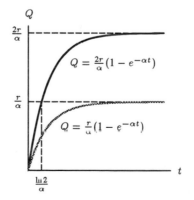

(c) Q_∞ is halved by doubling α, and so is the time, $t = \frac{\ln 2}{\alpha}$, to reach $\frac{1}{2}Q_\infty$.

17. (a) If $B = f(t)$ (where t is in years)

$$\frac{dB}{dt} = \text{(rate of money earned by interest)} + \text{(rate of money deposited)}$$

$$= 0.10B + 1000.$$

(b)
$$\frac{dB}{dt} = 0.1(B + 10000)$$
$$\int \frac{dB}{B + 10000} = \int 0.1 \, dt$$
$$\ln|B + 10000| = 0.1t + C$$
$$B = Ae^{0.1t} - 10000.$$

Since the initial balance is 0, then $B = 10000e^{0.1t} - 10000$.

(c) Suppose a deposit is made at time x. Then at time $t > x$, this deposit will have earned interest for $t - x$ years. We use this fact to set up the Riemann sum. Suppose we want to find the balance at time t. Divide it up into pieces of size Δx. The deposit made at time x is thus $\$1000 \cdot \Delta x$, and at time t it is worth $\$1000 \cdot \Delta x e^{0.1(t-x)}$. Thus our Riemann sum looks like $\sum 1000e^{0.1(t-x)}\Delta x$, and the corresponding integral is $f(t) = \int_0^t 1000e^{0.1(t-x)} \, dx$.

(d) $\int_0^t 1000e^{0.1(t-x)} \, dx = 1000[-10e^{0.1(t-x)}]\Big|_0^t = 10000e^{0.1t} - 10000$. This is the same answer as in part (b).

(e) If the initial deposit is B_0, then $B_0 = A - 10000$, so the solution is $B = (B_0 + 10000)e^{0.1t} - 10000$.

(f) The integral in part (d) actually stays the same. If the initial deposit is B_0, though, our integral doesn't take into account what happens to it. At time t, the initial deposit of B_0 is worth $B_0 e^{0.1t}$, so adding this in we would get

$$B = f(t) = B_0 e^{0.1t} + \int_0^t 1000e^{0.1(t-x)} \, dx$$

$$= B_0 e^{0.1t} + 10000e^{0.1t} - 10000$$

$$= (B_0 + 10000)e^{0.1t} - 10000.$$

This matches the answer of part (e).

21. (a) For a stable universe, we need $R' = 0$, so $R'' = 0$. However the differential equation for R'' shows that $R'' < 0$ for every R, so we never have $R'' = 0$. Thus R' and R must both be changing with time.

(b) If the universe were expanding at a constant rate of $R'(t_0)$, then $R(t_0)/R'(t_0)$ would be the time it took for the radius to grow from 0 to $R(t_0)$ – a reasonable estimate for the age of the universe. Since in fact R' has been decreasing, in other words, the universe has actually been expanding faster than $R'(t_0)$, the Hubble constant is an overestimate (i.e. the universe is actually younger than the Hubble constant suggests.)

9.7 SOLUTIONS

1. A continuous growth rate of 0.2% means that

$$\frac{1}{P}\frac{dP}{dt} = 0.2\% = 0.002.$$

Separating variables and integrating gives

$$\int \frac{dP}{P} = \int 0.002\, dt$$

$$P = P_0 e^{0.002t} = (6.6 \times 10^6)e^{0.002t}.$$

5. (a) Let I be the number of informed people at time t, and I_0 the number who know initially. Then this model predicts that $\frac{dI}{dt} = k(M - I)$ for some positive constant k. Solving this, we find the solution is

$$I = M - (M - I_0)e^{-kt}.$$

We sketch the solution with $I_0 = 0$. Notice that $\frac{dI}{dt}$ is largest when I is smallest, so the information spreads fastest in the beginning, at $t = 0$. In addition, the graph shows that $I \to M$ as $t \to \infty$, meaning that everyone gets the information eventually.

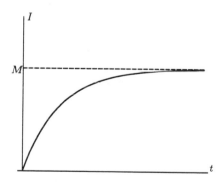

(b) In this case, the model suggests that $\frac{dI}{dt} = kI(M - I)$ for some positive constant k. This is a logistic model with carrying capacity M. We sketch the solutions for three different values of I_0.

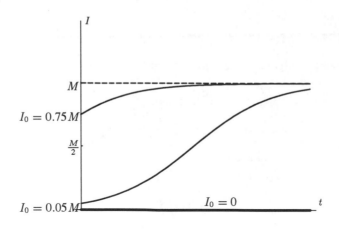

(i)   If $I_0 = 0$. Then $I = 0$ for all t. In other words, if nobody knows something, it doesn't spread by word of mouth!

(ii)   If $I_0 = 0.05 M$, then $\frac{dI}{dt}$ is increasing up to $I = \frac{M}{2}$. Thus, the information is spreading fastest at $I = \frac{M}{2}$.

(iii)   If $I_0 = 0.75 M$, then $\frac{dI}{dt}$ is always decreasing for $I > \frac{M}{2}$, so $\frac{dI}{dt}$ is largest when $t = 0$.

9.   (a)   Here we have, where $t =$ years since 1800:

TABLE 9.7

Year	t	$\frac{1}{P}\frac{dP}{dt}$
1800	0	0.0311
1830	30	0.0291
1860	60	0.0247
1890	90	0.0205
1920	120	0.0146
1950	150	0.0157
1980	180	0.0096

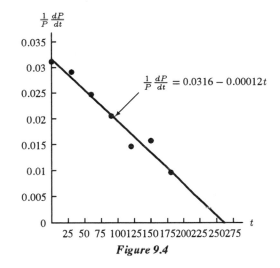

Figure 9.4

Graphing the data and fitting a line, we get $\frac{1}{P}\frac{dP}{dt} = 0.0316 - 0.00012t$ as our guess. So we have $a = 0.0316$ and $b = 0.00012$.

(b) $\frac{dP}{dt}$ will be positive and P will increase until $0.0316 = 0.00012t$, i.e. until $t \approx 260$ or about the year 2060.

(c) $\frac{dP}{P} = (0.0316 - 0.00012t)\,dt.$

$\int \frac{dP}{P} = \int(0.0316 - 0.00012t)\,dt.$

$P = Ae^{0.0316t - 0.00006t^2}.$

Using the fact that $P = 5.3$ when $t = 0$, we get $P = 5.3e^{0.0316t - 0.00006t^2}.$

13. If $f(x) = x^2$ then

$$\begin{aligned}
\frac{f(x+h) - f(x-h)}{2h} &= \frac{(x+h)^2 - (x-h)^2}{2h} \\
&= \frac{(x^2 + 2xh + h^2) - (x^2 - 2xh + h^2)}{2h} \\
&= \frac{4xh}{2h} = 2x = f'(x) \quad \text{for all } x.
\end{aligned}$$

17.   (a)

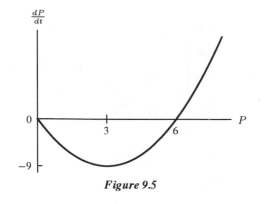

Figure 9.5

(b)   Figure 9.5 shows that for $0 < P < 6$, the sign of dP/dt is negative. This means that P is decreasing over the interval $0 < P < 6$. As P decreases from $P(0) = 5$, the value of dP/dt gets more and more negative until $P = 3$. Thus the graph of P against t is concave down while P is decreasing from 5 to 3. As P decreases below 3, the slope of dP/dt increases toward 0, so the graph of P against t is concave up and asymptotic to the t-axis. At $P = 3$, there is an inflection point. See Figure 9.6.

(c)   Figure 9.5 shows that for $P > 6$, the slope of dP/dt is positive and increases with P. Thus the graph of P against t is increasing and concave up. See Figure 9.6.

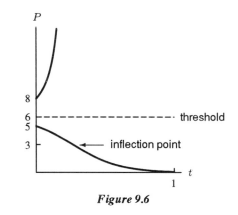

Figure 9.6

(d)   For initial populations greater than the threshold value $P = 6$, the population increases without bound. Populations with initial value less than $P = 6$ decrease asymptotically towards 0, i.e. become extinct. Thus the initial population $P = 6$ is the dividing line, or threshold, between populations which grow without bound and those which die out.

21. (a)

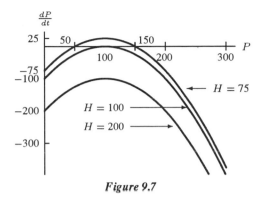

Figure 9.7

(b) For $H = 75$, the equilibrium populations (where $dP/dt = 0$) are $P = 50$ and $P = 150$. If the population is between 50 and 150, dP/dt is positive. This means that when the initial population is between 50 and 150, the population will increase until it reaches 150, when $dP/dt = 0$ and the population no longer increases or decreases. If the initial population is greater than 150, then dP/dt is negative, and the population decreases until it reaches 150. Thus 150 is a stable equilibrium. However, 50 is unstable.

For $H = 100$, the equilibrium population (where $dP/dt = 0$) is $P = 100$. For all other populations, dP/dt is negative and so the population decreases. If the initial population is greater than 100, it will decrease to the equilibrium value, $P = 100$. However, for populations less than 100, the population decreases until the species dies out.

For $H = 200$, there are no equilibrium points where $dP/dt = 0$, and dP/dt is always negative. Thus, no matter what the initial population, the population always dies out eventually.

(c) If the population is not to die out, looking at the three cases above, there must be an equilibrium value where $dP/dt = 0$, i.e. where the graph of dP/dt crosses the P axis. This happens if $H \leq 100$. Thus provided fishing is not more than 100 fish/year, there are initial values of the population for which the population will not be depleted.

(d) Fishing should be kept below the level of 100 per year.

9.8 SOLUTIONS

1. Since

$$\frac{dS}{dt} = -aSI,$$

$$\frac{dI}{dt} = aSI - bI,$$

$$\frac{dR}{dt} = bI$$

we have

$$\frac{dS}{dt} + \frac{dI}{dt} + \frac{dR}{dt} = -aSI + aSI - bI + bI = 0.$$

Thus $\frac{d}{dt}(S + I + R) = 0$, so $S + I + R = $ constant.

5. The closed trajectory represents populations which oscillate repeatedly.

9. If $w = 2$ and $r = 2$, then $\frac{dw}{dt} = -2$ and $\frac{dr}{dt} = 2$, so the number of worms decreases and the number of robins increases. In the long run, however, the populations will oscillate; they will even go back to $w = 2$ and $r = 2$.

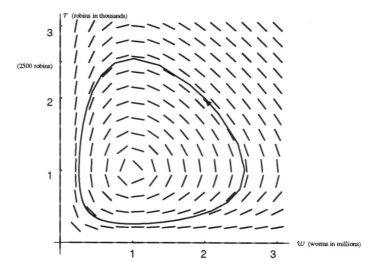

13. (a) Thinking of y as a function of x and x as a function of t, then by the chain rule: $\dfrac{dy}{dt} = \dfrac{dy}{dx}\dfrac{dx}{dt}$, so:

$$\frac{dy}{dx} = \frac{dy/dt}{dx/dt} = \frac{-0.01x}{-0.05y} = \frac{x}{5y}$$

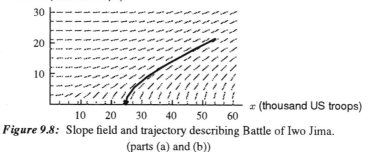

Figure 9.8: Slope field and trajectory describing Battle of Iwo Jima. (parts (a) and (b))

(b) Figure 9.8 shows the slope field for this differential equation and the trajectory starting at $x_0 = 54$, $y_0 = 21.5$. The trajectory goes to the x-axis, where $y = 0$, meaning that the Japanese troops were all killed or wounded before the US troops were, and thus predicts the US victory (which did occur). Since the trajectory meets the x-axis at $x \approx 25$, the differential equation predicts that about 25,000 US troops would survive the battle.

(c) The fact that the US got reinforcements, while the Japanese did not, does not alter the predicted outcome (a US victory). The US reinforcements have the effect of changing the trajectory, altering the number of troops surviving the battle. See the graph in Figure 9.9.

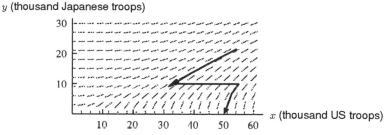

Figure 9.9: Trajectory describing Battle of Iwo Jima with US reinforcements

17. (a) If B were not present, then we'd have $A' = 2A$, so company A's net worth would grow exponentially. Similarly, if A were not present, B would grow exponentially. The two companies restrain each other's growth, probably by competing for the market.

(b) To find equilibrium points, find the solutions of the pair of equations

$$A' = 2A - AB = 0$$
$$B' = B - AB = 0$$

The first equation has solutions $A = 0$ or $B = 2$. The second has solutions $B = 0$ or $A = 1$. Thus the equilibrium points are (0,0) and (1,2).

(c) In the long run, one of the companies will go out of business. Two of the trajectories in the figure below go towards the A axis; they represent A surviving and B going out of business. The trajectories going towards the B axis represent A going out of business. Notice both the equilibrium points are unstable.

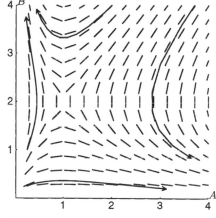

21. (a) Equilibrium points are where $\frac{dx}{dt} = 0$ and $\frac{dy}{dt} = 0$.
$15x - 3xy = 0$ gives $3x(5 - y) = 0$, so $x = 0$ or $y = 5$
$- 14y + 7xy = 0$ gives $7y(-2 + x) = 0$, so $x = 2$ or $y = 0$.
The solutions are thus (0,0) and (2,5).

(b)   At $x = 2$, $y = 0$ we have $\frac{dy}{dt} = 0$ but $\frac{dx}{dt} = 15(2) - 3(2)(0) = 30 \neq 0$. Thus $x = 2$, $y = 0$ is not an equilibrium point.

25.   (a)   Predator-prey, because x decreases while alone, but is helped by y, whereas y increases logistically when alone, and is harmed by x. Thus x is predator, y is prey.

(b)

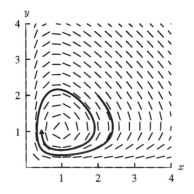

Provided neither initial population is zero, both populations tend to about 1. If x is initially zero, but y is not, then $y \to \infty$. If y is initially zero, but x is not, then $x \to 0$.

9.9   SOLUTIONS

1.   (a) $dS/dt = 0$ where $S = 0$ or $I = 0$ (both axes).
$dI/dt = 0.0026I(S - 192)$, so $dI/dt = 0$ where $I = 0$ or $S = 192$.
Thus every point on the S axis is an equilibrium point (corresponding to no one being sick).

(b)   In region I, where $S > 192$, $\dfrac{dS}{dt} < 0$ and $\dfrac{dI}{dt} > 0$.

In region II, where $S < 192$, $\dfrac{dS}{dt} < 0$ and $\dfrac{dI}{dt} < 0$. See Figure 9.10.

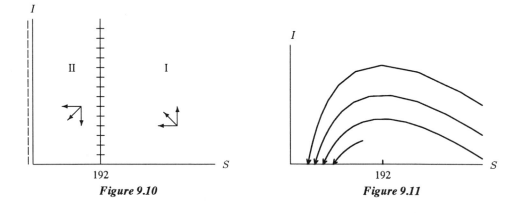

Figure 9.10   *Figure 9.11*

(c) If the trajectory starts with $S_0 > 192$, then I increases to a maximum when $S = 192$. If $S_0 < 192$, then I always decreases. See Figure 9.10. Regardless of the initial conditions, the trajectory always goes to a point on the S-axis (where $I = 0$). The S-intercept represents the number of students who never get the disease. See Figure 9.11.

5. We assume that x, $y \geq 0$ and then find the nullclines. $\frac{dx}{dt} = x(1 - \frac{x}{2} - y) = 0$ when $x = 0$ or $y + \frac{x}{2} = 1$. $\frac{dy}{dt} = y(1 - \frac{y}{3} - x) = 0$ when $y = 0$ or $x + \frac{y}{3} = 1$.

We find the equilibrium points. They are $(2, 0)$, $(0, 3)$, $(0, 0)$, and $(\frac{4}{5}, \frac{3}{5})$. The nullclines and equilibrium points are shown in Figure 9.12.

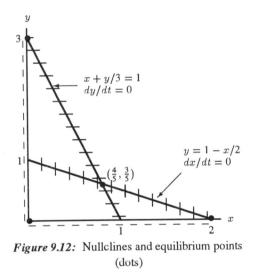

Figure 9.12: Nullclines and equilibrium points (dots)

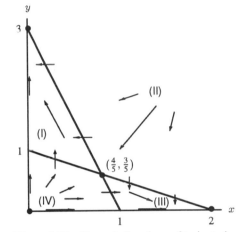

Figure 9.13: General directions of trajectories and equilibrium points (dots)

Figure 9.13 shows that if the initial point is in sector (I), the trajectory heads towards the equilibrium point $(0, 3)$. Similarly, if the trajectory begins in sector (III), then it heads towards the equilibrium $(2, 0)$ over time. If the trajectory begins in sector (II) or (IV), it can go to any of the three equilibrium points $(2, 0)$, $(0, 3)$, or $(\frac{4}{5}, \frac{3}{5})$.

9. (a) The nullclines are $P = 0$ or $P_1 + 3P_2 = 13$ (where $dP_1/dt = 0$) and $P = 0$ or $P_2 + 0.4P_1 = 6$ (where $dP_2/dt = 0$).

   (b) The phase plane in Figure 9.14 shows that P_2 will eventually exclude P_1 regardless of where the experiment starts so long as there were some P_2 originally. Consequently the data points would have followed a trajectory that starts at the origin, crosses the first nullcline and goes left and upwards between the two nullclines to the point $P_1 = 0$, $P_2 = 6$.

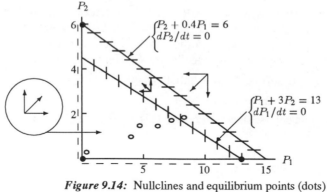

Figure 9.14: Nullclines and equilibrium points (dots)
for Gauses's yeast data (hollow dots)

9.10   SOLUTIONS

1. If $y = 2\cos t + 3\sin t$, then $y' = -2\sin t + 3\cos t$ and $y'' = -2\cos t - 3\sin t$. Thus, $y'' + y = 0$.

5. At $t = 0$, we find that $y = 2$, which is clearly the highest point since $-1 \le \cos 3t \le 1$. Thus, at $t = 0$ the mass is at its highest point. Since $y' = -6\sin 3t$, we see $y' = 0$ when $t = 0$. Thus, at $t = 0$ the object is at rest, although it will move down after $t = 0$.

9. First, we note that the solutions of:
    (a) $x'' + x = 0$ are $x = A\cos t + B\sin t$;
    (b) $x'' + 4x = 0$ are $x = A\cos 2t + B\sin 2t$;
    (c) $x'' + 16x = 0$ are $x = A\cos 4t + B\sin 4t$.
    This follows from what we know about the general solution to $x'' + \omega^2 x = 0$.
    The period of the solutions to (a) is 2π, the period of the solutions to (b) is π, and the period of the solutions of (c) is $\frac{\pi}{2}$.
    Since the t-scales are the same on all of the graphs, we see that graphs (I) and (IV) have the same period, which is twice the period of graph (III). Graph (II) has twice the period of graphs (I) and (IV). Since each graph represents a solution, we have the following:

    - equation (a) goes with graph (II)
      equation (b) goes with graphs (I) and (IV)
      equation (c) goes with graph (III)

    - The graph of (I) passes through $(0,0)$, so $0 = A\cos 0 + B\sin 0 = A$. Thus, the equation is $x = B\sin 2t$. Since the amplitude is 2, we see that $x = 2\sin 2t$ is the equation of the graph. Similarly, the equation for (IV) is $x = -3\sin 2t$.
      The graph of (II) also passes through $(0,0)$, so, similarly, the equation must be $x = B\sin t$. In this case, we see that $B = -1$, so $x = -\sin t$.
      Finally, the graph of (III) passes through $(0,1)$, and 1 is the maximum value. Thus, $1 = A\cos 0 + B\sin 0$, so $A = 1$. Since it reaches a local maximum at $(0,1)$, $x'(0) = 0 = -4A\sin 0 + 4B\cos 0$, so $B = 0$. Thus, the solution is $x = \cos 4t$.

13.   The amplitude is $\sqrt{3^2 + 7^2} = \sqrt{58}$.

17.   The equation we have for the charge tells us that:

$$\frac{d^2Q}{dt^2} = -\frac{Q}{LC},$$

where L and C are positive.

If we let $\omega = \sqrt{\frac{1}{LC}}$, we know the solution is of the form:

$$Q = C_1 \cos \omega t + C_2 \sin \omega t.$$

Since $Q(0) = 0$, we find that $C_1 = 0$, so $Q = C_2 \sin \omega t$.

Since $Q'(0) = 4$, and $Q' = \omega C_2 \cos \omega t$, we have $C_2 = \frac{4}{\omega}$, so $Q = \frac{4}{\omega} \sin \omega t$.

But we want the maximum charge, meaning the amplitude of Q, to be $2\sqrt{2}$ coulombs. Thus, we have $\frac{4}{\omega} = 2\sqrt{2}$, which gives us $\omega = \sqrt{2}$.

So we now have: $\sqrt{2} = \frac{1}{\sqrt{LC}} = \frac{1}{\sqrt{10C}}$. Thus, $C = \frac{1}{20}$ farads.

18.   We know that the general formula for Q will be of the form:

$$Q = C_1 \cos \omega t + C_2 \sin \omega t.$$

and

$$I = Q' = -C_1 \sin \omega t + C_2 \cos \omega t$$

Thus, as $t \to \infty$, neither one approaches a limit. Instead, they vary sinusoidally, with the same frequency but out of phase. We can think of the charge on the capacitor as being analogous to the displacement of a mass on a spring, oscillating from positive to negative. The current is then like the velocity of the mass, also oscillating from positive to negative. When the charge is maximal or minimal, the current is zero (just like when the spring is at the top or bottom of its motion), and when the current is maximal, the charge is zero (just like when the spring is at the middle of its motion).

9.11   SOLUTIONS

1.   (a)   The data in Table 9.17 on page 570 shows that each peak is 0.53 times the height of the previous one (since $0.5039/0.9445 \approx 0.53$ and $0.2668/0.5039 \approx 0.53$, etc.). Assuming that the peaks are a distance π apart, this means that every time t is increased by π, the value of the exponential function should decrease by a factor of 0.53. Writing $a^t = (0.53)^{t/\pi}$ achieves this.

     (b)   We can write $(0.53)^{t/\pi} = e^{kt}$ for some value of k. Solving for k:

$$\left((0.53)^{1/\pi}\right)^t = (e^k)^t$$

so

$$e^k = (0.53)^{1/\pi}.$$

Taking logs:

$$k = \frac{1}{\pi} \ln(0.53) = -0.20$$

Thus

$$(0.53)^{1/\pi} = e^{-0.20}$$

and

$$(0.53)^{t/\pi} = \left((0.53)^{1/\pi}\right)^t = (e^{-0.20})^t = e^{-0.20t}.$$

5.

s	v	$ds = v\Delta t$	$dv = -s\Delta t$
0	1	0.5	0
0.5	1	0.5	−0.25
1	0.75	0.375	−0.5

9.

$i = t$	S	I	dS/dt	dI/dt
0	762	1	−1.98	1.48
1	760.02	2.48	−4.90	3.66
2	755.12	6.14	−12.06	8.99
3	743.05	15.13	−29.24	21.67
4	713.82	36.80	−68.30	49.90
5	645.51	86.71	−145.52	102.17

9.12 SOLUTIONS

1. The characteristic equation is $r^2 + 4r + 3 = 0$, so $r = -1$ or -3.
   Therefore $y(t) = C_1 e^{-t} + C_2 e^{-3t}$.

5. The characteristic equation is $r^2 + 7 = 0$, so $r = \pm\sqrt{7}i$.
   Therefore $s(t) = C_1 \cos \sqrt{7}t + C_2 \sin \sqrt{7}t$.

9. The characteristic equation is $r^2 + 6r + 5 = 0$, so $r = -1$ or -5.
   Therefore $y(t) = C_1 e^{-t} + C_2 e^{-5t}$.
   $y'(t) = -C_1 e^{-t} - 5C_2 e^{-5t}$
   $y'(0) = 0 = -C_1 - 5C_2$
   $y(0) = 1 = C_1 + C_2$
   Therefore $C_2 = -1/4$, $C_1 = 5/4$ and $y(t) = \frac{5}{4}e^{-t} - \frac{1}{4}e^{-5t}$.

13. The characteristic equation is $r^2 + 2r + 2 = 0$, so $r = -1 \pm i$.
   Therefore $p(t) = C_1 e^{-t} \cos t + C_2 e^{-t} \sin t$.
   $p(0) = 0 = C_1$ so $p(t) = C_2 e^{-t} \sin t$
   $p(\pi/2) = 20 = C_2 e^{-\pi/2} \sin \frac{\pi}{2}$ so $C_2 = 20e^{\pi/2}$
   Therefore $p(t) = 20e^{\frac{\pi}{2}} e^{-t} \sin t = 20e^{\frac{\pi}{2}-t} \sin t$.

17. The restoring force is given by $F_{\text{spring}} = -ks$, so we look for the smallest coefficient of s. Spring (iv) exerts the smallest restoring force.

21. Recall that $s'' + bs' + c = 0$ is overdamped if the discriminant $b^2 - 4c > 0$, critically damped if $b^2 - 4c = 0$, and underdamped if $b^2 - 4c < 0$. Since $b^2 - 4c = 16 - 4c$, the circuit is overdamped if $c < 4$, critically damped if $c = 4$, and underdamped if $c > 4$.

25. The characteristic equation is $r^2 + r - 2 = 0$, so $r = 1$ or -2. Therefore $z(t) = C_1 e^t + C_2 e^{-2t}$. Since $e^t \to \infty$ as $t \to \infty$, we must have $C_1 = 0$. Therefore $z(t) = C_2 e^{-2t}$. Furthermore, $z(0) = 3 = C_2$, so $z(t) = 3e^{-2t}$.

29. (a) $\dfrac{d^2 y}{dt^2} = -\dfrac{dx}{dt} = y$ so $\dfrac{d^2 y}{dt^2} - y = 0.$

   (b) Characteristic equation $r^2 - 1 = 0$, so $r = \pm 1$.
     The general solution for y is $y = C_1 e^t + C_2 e^{-t}$, so $x = C_2 e^{-t} - C_1 e^t$.

33. The differential equation for the charge on the capacitor, given a resistance R, a capacitance C, and and inductance L, is

$$LQ'' + RQ' + \frac{Q}{C} = 0.$$

The corresponding characteristic equation is $Lr^2 + Rr + \dfrac{1}{C} = 0$. This equation has roots

$$r = -\frac{R}{2L} \pm \frac{\sqrt{R^2 - \frac{4L}{C}}}{2L}.$$

   (a) If $R^2 - \dfrac{4L}{C} < 0$, the solution is

$$Q(t) = e^{-\frac{R}{2L}t}(A \sin \omega t + B \cos \omega t) \text{ for some } A \text{ and } B,$$

     where $\omega = \dfrac{\sqrt{R^2 - \frac{4L}{C}}}{2L}$. As $t \to \infty$, $Q(t)$ clearly goes to 0.

   (b) If $R^2 - \dfrac{4L}{C} = 0$, the solution is

$$Q(t) = e^{-\frac{R}{t}}(A + Bt) \text{ for some } A \text{ and } B.$$

     Again, as $t \to \infty$, the charge goes to 0.

   (c) If $R^2 - \dfrac{4L}{C} > 0$, the solution is

$$Q(t) = Ae^{r_1 t} + Be^{r_2 t} \text{ for some } A \text{ and } B,$$

     where

$$r_1 = -\frac{R}{2L} + \frac{\sqrt{R^2 - \frac{4L}{C}}}{2L}, \quad \text{and} \quad r_2 = -\frac{R}{2L} - \frac{\sqrt{R^2 - \frac{4L}{C}}}{2L}.$$

Notice that r_2 is clearly negative. r_1 is also negative since

$$\frac{\sqrt{R^2 - \frac{4L}{C}}}{2L} < \frac{\sqrt{R^2}}{2L} \quad (L \text{ and } C \text{ are positive})$$
$$= \frac{R}{2L}.$$

Since r_1 and r_2 are negative, again $Q(t) \to 0$, as $t \to \infty$.

Thus, for any circuit with a resistor, a capacitor and an inductor, $Q(t) \to 0$ as $t \to \infty$. Compare this with Problem 18 in Section 9.10, where we showed that in a circuit with just a capacitor and inductor, the charge varied along a sine curve.

SOLUTIONS TO REVIEW PROBLEMS FOR CHAPTER NINE

1. $\frac{dP}{dt} = 0.03P + 400$ so $\int \frac{dP}{P + \frac{40000}{3}} = \int 0.03 dt$.

   $\ln |P + \frac{40000}{3}| = 0.03t + C$ giving $P = Ae^{0.03t} - \frac{40000}{3}$. Since $P(0) = 0, A = \frac{40000}{3}$, therefore $P = \frac{40000}{3}(e^{0.03t} - 1)$.

5. $\frac{dy}{dx} = e^{x-y}$ giving $\int e^y \, dy = \int e^x \, dx$ so $e^y = e^x + C$. Since $y(0) = 1$, we have $e^1 = e^0 + C$ so $C = e - 1$. Thus, $e^y = e^x + e - 1$, so $y = \ln(e^x + e - 1)$.
   [Note: $e^x + e - 1 > 0$ always.]

9. $\frac{dy}{dx} + xy^2 = 0$ means $\frac{dy}{dx} = -xy^2$, so $\int \frac{dy}{y^2} = \int -x \, dx$ giving $-\frac{1}{y} = -\frac{x^2}{2} + C$. Since $y(1) = 1$ we have $-1 = -\frac{1}{2} + C$ so $C = -\frac{1}{2}$. Thus, $-\frac{1}{y} = -\frac{x^2}{2} - \frac{1}{2}$ giving $y = \frac{2}{x^2+1}$.

13. $(1 + t^2)y \frac{dy}{dt} = 1 - y$ implies that $\int \frac{y \, dy}{1-y} = \int \frac{dt}{1+t^2}$ implies that $\int \left(-1 + \frac{1}{1-y}\right) dy = \int \frac{dt}{1+t^2}$. Therefore $-y - \ln|1 - y| = \arctan t + C$. $y(1) = 0$, so $0 = \arctan 1 + C$, and $C = -\frac{\pi}{4}$, so $-y - \ln|1 - y| = \arctan t - \frac{\pi}{4}$. We cannot solve for y in terms of t.

17. $(y\sqrt{x^3 + 1})\frac{dy}{dx} + x^2y^2 + x^2 = 0$ is equivalent to $(y\sqrt{x^3 + 1})\frac{dy}{dx} = -x^2y^2 - x^2 = -x^2(y^2 + 1)$. Separating variables yields $\frac{y \, dy}{y^2+1} = -\frac{x^2}{\sqrt{x^3+1}} dx$. Integrating, we obtain $\int \frac{y \, dy}{y^2+1} = -\int \frac{x^2}{\sqrt{x^3+1}} dx$.

   This implies $\frac{1}{2} \ln|y^2 + 1| = -\frac{2}{3}\sqrt{x^3 + 1} + C$, whence $y^2 + 1 = Ae^{-\frac{4}{3}\sqrt{x^3+1}}$ where $A = \pm e^{2C}$. So $y = \pm\sqrt{Ae^{-\frac{4}{3}\sqrt{x^3+1}} - 1}$.
   Note that A cannot be 0; in fact A must be greater than 1.

21. Using separation of variables and the integral tables, you can show that solutions are of the form $t = \frac{1}{4}(\ln|w - 7| - \ln|w - 3|) + C$, or $w = \frac{4}{1 - Ae^{4t}} + 3$. The equilibrium values, where $\frac{dw}{dt} = 0$, are $w = 3$ and $w = 7$. Graphs of the solutions can also be sketched directly from the graph of $\frac{dw}{dt}$ against w.

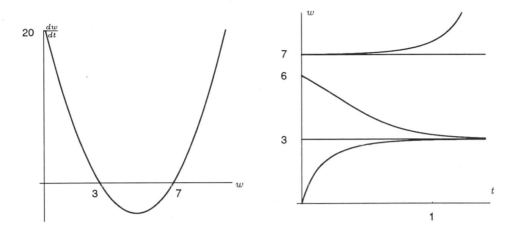

25. The characteristic equation of $9z'' + z = 0$ is

$$9r^2 + 1 = 0$$

If we write this in the form $r^2 + br + c = 0$, we have that $r^2 + 1/9 = 0$ and

$$b^2 - 4c = 0 - (4)(1/9) = -4/9 < 0$$

This indicates underdamped motion and since the roots of the characteristic equation are $r = \pm \frac{1}{3}i$, the general equation is

$$y(t) = C_1 \cos\left(\frac{1}{3}t\right) + C_2 \sin\left(\frac{1}{3}t\right)$$

29. Recall that $s'' + bs' + cs = 0$ is overdamped if the discriminant $b^2 - 4c > 0$, critically damped if $b^2 - 4c = 0$, and underdamped if $b^2 - 4c < 0$. This has discriminant $b^2 - 4c = b^2 + 64$. Since $b^2 + 64$ is always positive, the solution is always overdamped.

33. (a) For this situation,

$$\left(\begin{array}{c} \text{Rate money added} \\ \text{to account} \end{array}\right) = \left(\begin{array}{c} \text{Rate money added} \\ \text{via interest} \end{array}\right) + \left(\begin{array}{c} \text{Rate money} \\ \text{deposited} \end{array}\right)$$

Translating this into an equation yields

$$\frac{dB}{dt} = 0.1B + 1200$$

(b) Solving this equation via separation of variables gives

$$\frac{dB}{dt} = 0.1B + 1200$$

$$= (0.1)(B + 12000)$$

So

$$\int \frac{dB}{B + 12000} = \int 0.1 \, dt$$

and

$$\ln|B + 12000| = 0.1t + C$$

solving for B,

$$|B + 12000| = e^{(0.1)t + C} = e^C e^{(0.1)t}$$

or

$$B = Ae^{0.1t} - 12000, \ (\text{where } A = e^c)$$

We may find A using the initial condition $B_0 = f(0) = 0$

$$A - 12000 = 0 \quad \text{or} \quad A = 12000$$

So the solution is

$$B = f(t) = 12000(e^{0.1t} - 1)$$

(c)   After 5 years, the balance is

$$B = f(5) = 12000(e^{(0.1)(5)} - 1)$$
$$= 7784.66$$

37.   (a)   Since the guerrillas are hard to find, the rate at which they are put out of action is proportional to the number of chance encounters between a guerrilla and a conventional soldier, which is in turn proportional to the number of guerrillas and to the number of conventional soldiers. Thus the rate at which guerrillas are put out action is proportional to the product of the strengths of the two armies.

(b)
$$\frac{dx}{dt} = -xy$$
$$\frac{dy}{dt} = -x$$

(c)   Thinking of y as a function of x and x a function of of t, then by the chain rule: $\dfrac{dy}{dt} = \dfrac{dy}{dx}\dfrac{dx}{dt}$ so:

$$\frac{dy}{dx} = \frac{dy/dt}{dx/dt} = \frac{-x}{-xy} = \frac{1}{y}$$

Separating variables:

$$\int y \, dy = \int dx$$
$$\frac{y^2}{2} = x + C$$

The value of C is determined by the initial strengths of the two armies.

(d)   The sign of C determines which side wins the battle. Looking at the general solution $\frac{y^2}{2} = x + C$, we see that if $C > 0$ the y-intercept is at $\sqrt{2C}$, so y wins the battle by virtue of the fact that it still has troops when $x = 0$. If $C < 0$ then the curve intersects the axes at $x = -C$, so x wins the battle because it has troops when $y = 0$. If $C = 0$, then the solution goes to the point $(0, 0)$, which represents the case of mutual annihilation.

(e)   We assume that an army wins if the opposing force goes to 0 first. Figure 9.15 shows that the conventional force wins if $C > 0$ and the guerrillas win if $C < 0$. Neither side wins if $C = 0$ (all soldiers on both sides are killed in this case).

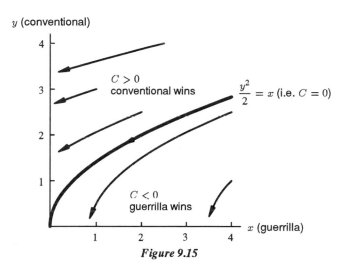

Figure 9.15

CHAPTER TEN

1. Let $f(x) = \cos x$. Then $f(0) = \cos(0) = 1$, and

$$
\begin{array}{ll}
f'(x) = -\sin x & f'(0) = 0 \\
f''(x) = -\cos x & f''(0) = -1 \\
f'''(x) = \sin x & f'''(0) = 0 \\
f^{(4)}(x) = \cos x & f^{(4)}(0) = 1 \\
f^{(5)}(x) = -\sin x & f^{(5)}(0) = 0 \\
f^{(6)}(x) = -\cos x & f^{(6)}(0) = -1.
\end{array}
$$

Thus,

$$
P_2(x) = 1 - \frac{x^2}{2!};
$$

$$
P_4(x) = 1 - \frac{x^2}{2!} + \frac{x^4}{4!};
$$

$$
P_6(x) = 1 - \frac{x^2}{2!} + \frac{x^4}{4!} - \frac{x^6}{6!}.
$$

5. Let $f(x) = \tan x$. So $f(0) = \tan 0 = 0$, and

$$
\begin{array}{ll}
f'(x) = 1/\cos^2 x & f'(0) = 1, \\
f''(x) = 2\sin x/\cos^3 x & f''(0) = 0, \\
f'''(x) = (2/\cos^2 x) + (6\sin^2 x/\cos^4 x) & f'''(0) = 2, \\
f^{(4)}(x) = (16\sin x/\cos^3 x) + (24\sin^3 x/\cos^5 x) & f^{(4)}(0) = 0.
\end{array}
$$

Thus,

$$
P_3(x) = P_4(x) = x + \frac{x^3}{3}.
$$

9. Let $f(x) = \dfrac{1}{\sqrt{1+x}} = (1+x)^{-1/2}$. Then $f(0) = 1$

$$
\begin{array}{ll}
f'(x) = -\frac{1}{2}(1+x)^{-3/2} & f'(0) = -\frac{1}{2}, \\
f''(x) = \frac{3}{2^2}(1+x)^{-5/2} & f''(0) = \frac{3}{2^2}, \\
f'''(x) = -\frac{3\cdot5}{2^3}(1+x)^{-7/2} & f'''(0) = -\frac{3\cdot5}{2^3}, \\
f^{(4)}(x) = \frac{3\cdot5\cdot7}{2^4}(1+x)^{-9/2} & f^{(4)}(0) = \frac{3\cdot5\cdot7}{2^4}.
\end{array}
$$

Then,

$$
P_2(x) = 1 - \frac{1}{2}x + \frac{1}{2!}\frac{3}{2^2}x^2 = 1 - \frac{1}{2}x + \frac{3}{8}x^2,
$$

$$P_3(x) = P_2(x) - \frac{1}{3!}\frac{3 \cdot 5}{2^3}x^3 = 1 - \frac{1}{2}x + \frac{3}{8}x^2 - \frac{5}{16}x^3,$$

$$P_4(x) = P_3(x) + \frac{1}{4!}\frac{3 \cdot 5 \cdot 7}{2^4}x^4 = 1 - \frac{1}{2}x + \frac{3}{8}x^2 - \frac{5}{16}x^3 + \frac{35}{128}x^4.$$

13. Let $f(x) = \cos x$. $f(\frac{\pi}{2}) = 0$.

$$
\begin{array}{ll}
f'(x) = -\sin x & f'(\frac{\pi}{2}) = -1, \\
f''(x) = -\cos x & f''(\frac{\pi}{2}) = 0, \\
f'''(x) = \sin x & f'''(\frac{\pi}{2}) = 1, \\
f^{(4)}(x) = \cos x & f^{(4)}(\frac{\pi}{2}) = 0.
\end{array}
$$

So,

$$P_4(x) = 0 - \left(x - \frac{\pi}{2}\right) + 0 + \frac{1}{3!}\left(x - \frac{\pi}{2}\right)^3$$

$$= -\left(x - \frac{\pi}{2}\right) + \frac{1}{3!}\left(x - \frac{\pi}{2}\right)^3.$$

17. Let $f(x) = \sqrt{1-x} = (1-x)^{1/2}$. Then $f'(x) = -\frac{1}{2}(1-x)^{-1/2}$, $f''(x) = -\frac{1}{4}(1-x)^{-3/2}$, $f'''(x) = -\frac{3}{8}(1-x)^{-5/2}$. So $f(0) = 1$, $f'(0) = -\frac{1}{2}$, $f''(0) = -\frac{1}{4}$, $f'''(0) = -\frac{3}{8}$, and

$$P_3(x) = 1 - \frac{1}{2}x - \frac{1}{4}\frac{1}{2!}x^2 - \frac{3}{8}\frac{1}{3!}x^3$$

$$= 1 - \frac{x}{2} - \frac{x^2}{8} - \frac{x^3}{16}.$$

21. This is the same as Example 8, Page 597 except we need two more terms:

$$
\begin{array}{ll}
f^{(5)}(x) = 24x^{-5} & f^{(5)}(1) = 24, \\
f^{(6)}(x) = -120x^{-6} & f^{(6)}(1) = -120.
\end{array}
$$

So,

$$P_6(x) = P_4(x) + \frac{24}{5!}(x-1)^5 + \frac{-120}{6!}(x-1)^6$$

$$= (x-1) - \frac{(x-1)^2}{2} + \frac{(x-1)^3}{3} - \frac{(x-1)^4}{4} + \frac{(x-1)^5}{5} - \frac{(x-1)^6}{6}.$$

23. Since $P_2(x)$ is the second degree Taylor polynomial for $f(x)$ about $x = 0$, $P_2(0) = f(0)$, which says $a = f(0)$; Since

$$\left.\frac{d}{dx}P_2(x)\right|_{x=0} = f'(0),$$

$b = f'(0)$; and since

$$\left.\frac{d^2}{dx^2} P_2(x)\right|_{x=0} = f''(0),$$

$2c = f''(0)$. As we can see now, a is the y-intercept of $f(x)$, b is the slope of the tangent line to $f(x)$ at $x = 0$ and c tells us the concavity of $f(x)$ near $x = 0$. So $c < 0$ since f is concave down; $b > 0$ since f is increasing; $a > 0$ since $f(0) > 0$.

25. As we can see from Problem 23, a is the y-intercept of $f(x)$, b is the slope of the tangent line to $f(x)$ at $x = 0$ and c tells us the concavity of $f(x)$ near $x = 0$.
So $a < 0$, $b > 0$ and $c > 0$.

29. For $f(h) = e^h$, $P_4(h) = 1 + h + \dfrac{h^2}{2} + \dfrac{h^3}{3!} + \dfrac{h^4}{4!}$. So

(a)

$$\begin{aligned}
\lim_{h \to 0} \frac{e^h - 1 - h}{h^2} &= \lim_{h \to 0} \frac{e^h - 1 - h}{h^2} \\
&= \lim_{h \to 0} \frac{\frac{h^2}{2} + \frac{h^3}{3!} + \frac{h^4}{4!}}{h^2} = \lim_{h \to 0} \left(\frac{1}{2} + \frac{h}{3!} + \frac{h^2}{4!} \right) \\
&= \frac{1}{2}.
\end{aligned}$$

(b)

$$\begin{aligned}
\lim_{h \to 0} \frac{e^h - 1 - h - \frac{h^2}{2}}{h^3} &= \lim_{h \to 0} \frac{P_4(h) - 1 - h - \frac{h^2}{2}}{h^3} \\
&= \lim_{h \to 0} \frac{\frac{h^3}{3!} + \frac{h^4}{4!}}{h^3} = \lim_{h \to 0} \left(\frac{1}{3!} + \frac{h}{4!} \right) \\
&= \frac{1}{3!} = \frac{1}{6}.
\end{aligned}$$

Using Taylor polynomials of higher degree would not have changed the results since the terms with higher powers of h all go to zero as $h \to 0$.

33. (a) We'll make the following conjecture:
"If $f(x)$ is a polynomial of degree n, i.e.

$$f(x) = a_0 + a_1 x + a_2 x^2 + \cdots + a_{n-1} x^{n-1} + a_n x^n,$$

then $P_n(x)$, the nth degree Taylor polynomial for $f(x)$ about $x = 0$, is $f(x)$ itself."

(b) All we need to do is to calculate $P_n(x)$, the nth degree Taylor polynomial for f about $x = 0$ and see if it is the same as $f(x)$.

$$C_0 = f(0) = a_0;$$

$$C_1 = f'(0) = (a_1 + 2a_2x + \cdots + na_nx^{n-1})\big|_{x=0}$$
$$= a_1;$$
$$C_2 = f''(0) = (2a_2 + 3 \cdot 2a_3x + \cdots + n(n-1)a_nx^{n-2})\big|_{x=0}$$
$$= 2!a_2.$$

If we continue doing this, we'll see in general

$$C_k = f^{(k)}(0) = k!a_k, \qquad k = 1, 2, 3, \cdots, n.$$

So, $a_k = \dfrac{C_k}{k!}$, $k = 1, 2, 3, \cdots, n$. Therefore,

$$P_n(x) = C_0 + \frac{C_1}{1!}x + \frac{C_2}{2!}x^2 + \cdots + \frac{C_n}{n!}x^n$$
$$= a_0 + a_1x + a_2x^2 + \cdots + a_nx^n$$
$$= f(x).$$

10.2 SOLUTIONS

1. Yes.

5. Yes.

9.

$$\begin{array}{ll}
f(t) = \ln(1-t) & f(0) = 0 \\
f'(t) = -\frac{1}{1-t} = -(1-t)^{-1} & f'(0) = -1 \\
f''(t) = (1-t)^{-2}(-1) = -(1-t)^{-2} & f''(0) = -1 \\
f'''(t) = 2(1-t)^{-3}(-1) = -2(1-t)^{-3} & f'''(0) = -2
\end{array}$$

$$f(t) = \ln(1-t) = -t - \frac{t^2}{2!} - \frac{2t^3}{3!} - \cdots$$

12.

$$\begin{array}{ll}
f(x) = \frac{1}{x} & f(1) = 1 \\
f'(x) = -\frac{1}{x^2} & f'(1) = -1 \\
f''(x) = \frac{2}{x^3} & f''(1) = 2 \\
f'''(x) = -\frac{6}{x^4} & f'''(1) = -6
\end{array}$$

$$\frac{1}{x} = 1 - (x-1) + \frac{2(x-1)^2}{2!} - \frac{6(x-1)^3}{3!} + \cdots$$
$$= 1 - (x-1) + (x-1)^2 - (x-1)^3 + \cdots.$$

13. Using the derivatives from Problem 12, we have

$$f(-1) = -1, \quad f'(-1) = -1, \quad f''(-1) = -2, \quad f'''(-1) = -6.$$

Hence,

$$\frac{1}{x} = -1 - (x+1) - \frac{2(x+1)^2}{2!} - \frac{6(x+1)^3}{3!} - \cdots$$
$$= -1 - (x+1) - (x+1)^2 - (x+1)^3 - \cdots$$

17.

$$
\begin{aligned}
f(\theta) &= \sin\theta & f(-\tfrac{\pi}{4}) &= -\tfrac{\sqrt{2}}{2} \\
f'(\theta) &= \cos\theta & f'(-\tfrac{\pi}{4}) &= \tfrac{\sqrt{2}}{2} \\
f''(\theta) &= -\sin\theta & f''(-\tfrac{\pi}{4}) &= \tfrac{\sqrt{2}}{2} \\
f'''(\theta) &= -\cos\theta & f'''(-\tfrac{\pi}{4}) &= -\tfrac{\sqrt{2}}{2}
\end{aligned}
$$

$$\sin\theta = -\frac{\sqrt{2}}{2} + \frac{\sqrt{2}}{2}\left(\theta + \frac{\pi}{4}\right) + \frac{\sqrt{2}}{2}\frac{(\theta + \frac{\pi}{4})^2}{2!} - \frac{\sqrt{2}}{2}\frac{(\theta + \frac{\pi}{4})^3}{3!} + \cdots$$
$$= -\frac{\sqrt{2}}{2} + \frac{\sqrt{2}}{2}\left(\theta + \frac{\pi}{4}\right) + \frac{\sqrt{2}}{4}\left(\theta + \frac{\pi}{4}\right)^2 - \frac{\sqrt{2}}{12}\left(\theta + \frac{\pi}{4}\right)^3 + \cdots.$$

21. Using the Binomial series with $p = 1/2$, we have

$$\sqrt{1+x} = (1+x)^{1/2} = 1 + \frac{x}{2} - \frac{x^2}{8} + \cdots.$$

Hence

$$
\begin{aligned}
\lim_{x \to 0} \frac{\sqrt{1+x} - 1}{x} &= \lim_{x \to 0} \frac{\left(1 + \frac{x}{2} - \frac{x^2}{8} + \cdots\right) - 1}{x} \\
&= \lim_{x \to 0} \left(\frac{1}{2} - \frac{x}{8} + \cdots\right) \\
&= \frac{1}{2}.
\end{aligned}
$$

25. (a)
$$
\begin{aligned}
f(x) &= \ln(1 + 2x) & f(0) &= 0 \\
f'(x) &= \frac{2}{1+2x} & f'(0) &= 2 \\
f''(x) &= -\frac{4}{(1+2x)^2} & f''(0) &= -4 \\
f'''(x) &= \frac{16}{(1+2x)^3} & f'''(0) &= 16
\end{aligned}
$$

$$\ln(1 + 2x) = 2x - 2x^2 + \frac{8}{3}x^3 + \cdots$$

(b) To get the expression for $\ln(1+2x)$ from the series for $\ln(1+x)$, substitute $2x$ for x in the series

$$\ln(1+x) = x - \frac{x^2}{2} + \frac{x^3}{3} - \frac{x^4}{4} + \cdots$$

to get

$$\ln(1+2x) = 2x - \frac{(2x)^2}{2} + \frac{(2x)^3}{3} - \frac{(2x)^4}{4} + \cdots$$

$$= 2x - 2x^2 + \frac{8x^3}{3} - 4x^4 + \cdots$$

(c) Since the interval of convergence for $\ln(1+x)$ is $-1 < x < 1$, substituting $2x$ for x suggests the interval of convergence of $\ln(1+2x)$ is $-1 < 2x < 1$ or $-\frac{1}{2} < x < \frac{1}{2}$

29. By looking at the graph we can see that the Taylor polynomials are reasonable approximations for the function $f(x) = \frac{1}{\sqrt{1+x}}$ between $x = -0.25$ and $x = 0.25$. Thus a good guess is that the interval of convergence is $-0.25 < x < 0.25$.

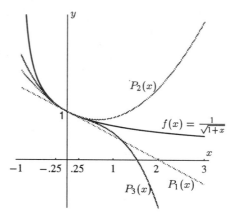

10.3 SOLUTIONS

1. Substitute $y = -x$ into $e^y = 1 + y + \frac{y^2}{2!} + \frac{y^3}{3!} + \cdots$. We get

$$e^{-x} = 1 + (-x) + \frac{(-x)^2}{2!} + \frac{(-x)^3}{3!} + \cdots$$

$$= 1 - x + \frac{x^2}{2!} - \frac{x^3}{3!} + \cdots$$

5.

$$\frac{t}{1+t} = t(1+t)^{-1} = t\left(1 + (-1)t + \frac{(-1)(-2)}{2!}t^2 + \frac{(-1)(-2)(-3)}{3!}t^3 + \cdots\right)$$

$$= t - t^2 + t^3 - t^4 + \cdots$$

9.

$$\phi^3 \cos(\phi^2) = \phi^3 \left(1 - \frac{(\phi^2)^2}{2!} + \frac{(\phi^2)^4}{4!} - \frac{(\phi^2)^6}{6!} + \cdots \right)$$

$$= \phi^3 - \frac{\phi^7}{2!} + \frac{\phi^{11}}{4!} - \frac{\phi^{15}}{6!} + \cdots$$

13.

$$\frac{1}{2+x} = \frac{1}{2(1+\frac{x}{2})} = \frac{1}{2}\left(1+\frac{x}{2}\right)^{-1}$$

$$= \frac{1}{2}\left(1 - \frac{x}{2} + \left(\frac{x}{2}\right)^2 - \left(\frac{x}{2}\right)^3 + \cdots\right)$$

17.   The Taylor series about 0 for $y = \dfrac{1}{1-x^2}$ is

$$y = 1 + x^2 + x^4 + x^6 + \cdots.$$

The series for $y = (1+x)^{\frac{1}{4}}$ is, using the binomial expansion,

$$y = 1 + \frac{1}{4}x + \frac{1}{4}\left(-\frac{3}{4}\right)\frac{x^2}{2!} + \frac{1}{4}\left(-\frac{3}{4}\right)\left(-\frac{7}{4}\right)\frac{x^3}{3!} + \cdots.$$

The series for $y = \sqrt{1+\dfrac{x}{2}} = (1+\dfrac{x}{2})^{\frac{1}{2}}$ is, again using the binomial expansion,

$$y = 1 + \frac{1}{2}\cdot\frac{x}{2} + \frac{1}{2}\left(-\frac{1}{2}\right)\cdot\frac{x^2}{8} + \frac{1}{2}\left(-\frac{1}{2}\right)\left(-\frac{3}{2}\right)\cdot\frac{x^3}{48} + \cdots.$$

Similarly for $y = \dfrac{1}{\sqrt{1-x}} = (1-x)^{-\frac{1}{2}}$,

$$y = 1 + \left(-\frac{1}{2}\right)(-x) + \left(-\frac{1}{2}\right)\left(-\frac{3}{2}\right)\cdot\frac{x^2}{2!} + \left(-\frac{1}{2}\right)\left(-\frac{3}{2}\right)\left(-\frac{5}{2}\right)\cdot\frac{-x^3}{3!} + \cdots.$$

Near 0, let's truncate these series after their x^2 terms:

$$\frac{1}{1-x^2} \approx 1 + x^2,$$

$$(1+x)^{\frac{1}{4}} \approx 1 + \frac{1}{4}x - \frac{3}{32}x^2,$$

$$\sqrt{1+\frac{x}{2}} \approx 1 + \frac{1}{4}x - \frac{1}{32}x^2,$$

$$\frac{1}{\sqrt{1-x}} \approx 1 + \frac{1}{2}x + \frac{3}{8}x^2.$$

Thus $\frac{1}{1-x^2}$ looks like a parabola opening upward near the origin, with y-axis as the axis of symmetry, so (a) = I.

Now $\frac{1}{\sqrt{1-x}}$ has the largest positive slope ($\frac{1}{2}$), and is concave up (because the coefficient of x^2 is positive). So (d) = II.

The last two both have positive slope ($\frac{1}{4}$) and are concave down. Since $(1+x)^{\frac{1}{4}}$ has the smallest second derivative (i.e. the most negative coefficient of x^2), (b) = IV and therefore (c) = III.

21.

$$E = kQ\left(\frac{1}{(R-1)^2} - \frac{1}{(R+1)^2}\right)$$

$$= \frac{kQ}{R^2}\left(\frac{1}{(1-\frac{1}{R})^2} - \frac{1}{(1+\frac{1}{R})^2}\right)$$

Since $|\frac{1}{R}| < 1$, we can expand the two terms using the binomial expansion:

$$\frac{1}{(1-\frac{1}{R})^2} = \left(1 - \frac{1}{R}\right)^{-2}$$

$$= 1 - 2\left(-\frac{1}{R}\right) + (-2)(-3)\frac{(-\frac{1}{R})^2}{2!} + (-2)(-3)(-4)\frac{(-\frac{1}{R})^3}{3!} + \cdots$$

$$\frac{1}{(1+\frac{1}{R})^2} = \left(1 + \frac{1}{R}\right)^{-2}$$

$$= 1 - 2\left(\frac{1}{R}\right) + (-2)(-3)\frac{(\frac{1}{R})^2}{2!} + (-2)(-3)(-4)\frac{(\frac{1}{R})^3}{3!} + \cdots$$

Substituting, we get:

$$E = \frac{kQ}{R^2}\left[1 + \frac{2}{R} + \frac{3}{R^2} + \frac{4}{R^3} + \cdots - \left(1 - \frac{2}{R} + \frac{3}{R^2} - \frac{4}{R^3} + \cdots\right)\right]$$

$$\approx \frac{kQ}{R^2}\left(\frac{4}{R} + \frac{8}{R^3}\right)$$

using only the first two non-zero terms.

25. (a) At $r = a$, the force between the atoms is 0.

(b)   There will be an attractive force, pulling them back together.

(c)   There will be a repulsive force, pushing the atoms apart.

(d) $F = F(a) + F'(a)(r - a) + F''(a)\frac{(r-a)^2}{2!} + \cdots$

(e) $F(a) = 0$, so discarding all but the first non-zero terms, we get $F = F'(a)(r - a)$. $F'(a)$ is a negative number, so for r slightly greater than a, the force is negative (attractive). For r slightly less than a, the force is positive (repulsive).

29.   Since $C_0 = f(0)$, the initial conditions $f(0) = 0$, $f'(0) = 1$ give $C_0 = 0$, $C_1 = 1$, so

$$f(\theta) = \theta + C_2\theta^2 + C_3\theta^3 + C_4\theta^4 + C_5\theta^5 + \cdots.$$

Taking the derivative of the series for f term by term twice, we get

$$\frac{df}{d\theta} = 1 + 2C_2\theta + 3C_3\theta^2 + 4C_4\theta^3 + 5C_5\theta^4 + \cdots.$$

$$\frac{d^2f}{d\theta^2} = 2 \cdot 1C_2 + 3 \cdot 2C_3\theta + 4 \cdot 3C_4\theta^2 + 5 \cdot 4C_5\theta^3 + \cdots.$$

Since the differential equation tells us that $\frac{d^2f}{d\theta^2} = -f$, we must have

$$2 \cdot 1C_2 + 3 \cdot 2C_3\theta + 4 \cdot 3C_4\theta^2 + 5 \cdot 4C_5\theta^3 + \cdots = -\theta - C_2\theta^2 - C_3\theta^3 - C_4\theta^4 \cdots.$$

Equate coefficients of corresponding powers of θ:

Constant terms	$2C_2 = 0$ so $C_2 = 0$
Coefficients of θ	$6C_3 = -C_1 = -1$, so $C_3 = -1/6$
Coefficients of θ^2	$12C_4 = -C_2 = 0$, so $C_4 = 0$
Coefficients of θ^3	$20C_5 = -C_3 = 1/6$ so $C_5 = 1/120$.

We have found the approximation

$$f(\theta) \approx \theta - \frac{\theta^3}{6} + \frac{\theta^5}{120} \quad \text{for } \theta \text{ near 0.}$$

In this case, we can see that the coefficients can be written as

$$C_3 = -\frac{1}{6} = -\frac{1}{3!}, \quad C_5 = \frac{1}{120} = \frac{1}{5!},$$

so we recognize the series as the Taylor series for $\sin\theta$ about $\theta = 0$.

10.4   SOLUTIONS

1.   Yes, $a = 2$, ratio $= 1/2$.

5.   Yes, $a = 1$, ratio $= -x$.

9.   Yes, $a = 1$, ratio $= 2z$.

13. Sum $= \dfrac{1}{1 - (-y^2)} = \dfrac{1}{1 + y^2}$, $|y| < 1$.

17.

$$\sum_{n=4}^{\infty} \left(\frac{1}{3}\right)^n = \left(\frac{1}{3}\right)^4 + \left(\frac{1}{3}\right)^5 + \cdots = \left(\frac{1}{3}\right)^4 \left(1 + \frac{1}{3} + \left(\frac{1}{3}\right)^2 + \cdots\right) = \frac{\left(\frac{1}{3}\right)^4}{1 - \frac{1}{3}} = \frac{1}{54}$$

21. (a)

$$P_1 = 0$$
$$P_2 = 250(0.04)$$
$$P_3 = 250(0.04) + 250(0.04)^2$$
$$P_4 = 250(0.04) + 250(0.04)^2 + 250(0.04)^3$$
$$\cdots$$
$$P_n = 250(0.04) + 250(0.04)^2 + 250(0.04)^3 + \cdots + 250(0.04)^{n-1}$$

(b)

$$P_n = 250(0.04)\left(1 + (0.04) + (0.04)^2 + (0.04)^3 + \cdots + (0.04)^{n-2}\right)$$
$$= 250\frac{0.04(1 - (0.04)^{n-1})}{1 - 0.04}$$

(c)

$$P = \lim_{n \to \infty} P_n$$
$$= \lim_{n \to \infty} 250\frac{0.04(1 - (0.04)^{n-1})}{1 - 0.04}$$
$$= \frac{(250)(0.04)}{0.96}$$
$$= 0.04Q$$
$$\approx 10.42$$

Thus, $\lim_{n \to \infty} P_n = 10.42$ and $\lim_{n \to \infty} Q_n = 260.42$. We'd expect these limits to differ because one is right before taking a tablet, one is right after. We'd expect the difference between them to be 250 mg, the amount of ampicillin in one tablet.

25. As in the text,

Present value of first payment, in millions of dollars $= 3$

Since the second payment is made a year in the future, so with continuous compounding,

Present value of second payment, in millions of dollars $= 3e^{-0.07}$

Since the next payment is two years in the future,

Present value of third payment, in millions of dollars $= 3e^{-0.07(2)}$

Similarly,

$$\text{Present value of tenth payment, in millions of dollars } = 3e^{-0.07(9)}$$

Thus, in millions of dollars,

$$\text{Total present value} = 3 + 3e^{-0.07} + 3e^{-0.07(2)} + 3e^{-0.07(3)} + \cdots + 3e^{-0.07(9)}$$

Since $e^{-0.07(n)} = \left(e^{-0.07}\right)^n$ for any n, we can write

$$\text{Total present value} = 3 + 3e^{-0.07} + 3\left(e^{-0.07}\right)^2 + 3\left(e^{-0.07}\right)^3 + 3\left(e^{-0.07}\right)^9.$$

This is a finite geometric series with $x = e^{-0.07}$ and sum

$$\text{Total present value of contract, in millions of dollars} = \frac{3\left(1 - (e^{-0.07})^{10}\right)}{1 - e^{-0.07}} \approx 22.3$$

29. A person should expect to pay the present value of the bond on the day it is bought.

$$\text{Present value of first payment } = \frac{10}{1.04}$$
$$\text{Present value of second payment } = \frac{10}{(1.04)^2} \text{ and so on.}$$

Therefore,

$$\text{Total present value } = \frac{10}{1.04} + \frac{10}{(1.04)^2} + \frac{10}{(1.04)^3} + \cdots$$

This is a geometric series with $a = \frac{10}{1.04}$ and $x = \frac{1}{1.04}$, so

$$\text{Total present value } = \frac{\frac{10}{1.04}}{1 - \frac{1}{1.04}} = £250.$$

33. (a)

$$\text{Total amount of money deposited} = 100 + 92 + 84.64 + \cdots$$
$$= 100 + 100(0.92) + 100(0.92)^2 + \cdots$$
$$= \frac{100}{1 - 0.92} = 1250 \text{ dollars}$$

(b) Credit multiplier $= 1250/100 = 12.50$
The 12.50 is the factor by which the bank has increased its deposits, from $100 to $1250.

10.5 SOLUTIONS

1. (a) The Taylor polynomial of degree 0 about $t = 0$ for $f(t) = e^t$ is simply $f(0) = 1$. Since $e^t \geq 1$ on $[0, 0.5]$, the approximation is an underestimate.

    (b) Using the zero degree error bound, if $|f'(t)| \leq M$ for $0 \leq t \leq 0.5$, then

    $$|E_0| \leq M \cdot |t| \leq M(0.5).$$

    Since $|f'(t)| = |e^t| = e^t$ is increasing on $[0, 0.5]$,

    $$|f'(t)| \leq e^{0.5} < \sqrt{3} \approx 1.732.$$

    Therefore

    $$|E_0| \leq (1.732)(0.5) = 0.8666.$$

    (Note: By looking at a graph of $f(t)$ and its 0th degree approximation, it is easy to see that the greatest error occurs when $t = 0.5$, and the error is $e^{0.5} - 1 \approx 0.65 < 0.866$. So our error bound works.)

5. Let $f(x) = \tan x$. The error bound for the Taylor approximation of degree three for $f(1) = \tan 1$ about $x = 0$ is:

    $$|E_3| = |f(1) - P_3(x)| \leq \frac{M \cdot |1 - 0|^4}{4!} = \frac{M}{24}$$

    where $|f^{(4)}(x)| \leq M$ for $0 \leq x \leq 1$. Now, $f^{(4)}(x) = \frac{16 \sin x}{\cos^3 x} + \frac{24 \sin^3 x}{\cos^5 x}$. From a graph of $f^{(4)}(x)$, we see that $f^{(4)}(x)$ is increasing for x between 0 and 1. Thus,

    $$|f^{(4)}(x)| \leq |f^{(4)}(1)| \approx 396,$$

    so

    $$|E_3| \leq \frac{396}{24} = 16.5.$$

    This is not a very helpful error bound! The reason the error bound is so huge is that $x = 1$ is getting near the vertical asymptote of the tangent graph, and the fourth derivative is enormous there.

9. The maximum possible error for the nth degree Taylor polynomial about $x = 0$ approximating $\cos x$ is $|E_n| \leq \frac{M \cdot |x - 0|^{n+1}}{(n+1)!}$, where $|\cos^{(n+1)} x| \leq M$ for $0 \leq x \leq 1$. Now the derivatives of $\cos x$ are simply $\cos x, \sin x, -\cos x$, and $-\sin x$. The largest magnitude these ever take is 1, so $|\cos^{(n+1)}(x)| \leq 1$, and thus $|E_n| \leq \frac{|x|^{n+1}}{(n+1)!} \leq \frac{1}{(n+1)!}$. The same argument works for $\sin x$.

13. (a)

TABLE 10.1 $E_1 = \sin x - x$

x	$\sin x$	E
-0.5	-0.4794	0.0206
-0.4	-0.3894	0.0106
-0.3	-0.2955	0.0045
-0.2	-0.1987	0.0013
-0.1	-0.0998	0.0002

TABLE 10.2 $E_1 = \sin x - x$

x	$\sin x$	E
0	0	0
0.1	0.0998	-0.0002
0.2	0.1987	-0.0013
0.3	0.2955	-0.0045
0.4	0.3894	-0.0106
0.5	0.4794	-0.0206

(b) See answer to part (a) above.

(c)

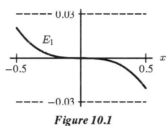

Figure 10.1

The fact that the graph of E_1 lies between the horizontal lines at $+0.03$ in Figure 10.1 shows that $|E_1| < 0.03$ for $-0.5 \leq x \leq 0.5$.

10.6 SOLUTIONS

1. Yes.

5. (a) (i) The graph of $y = \sin x + \frac{1}{3}\sin 3x$ looks like

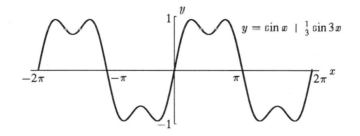

(ii) The graph of $y = \sin x + \frac{1}{3}\sin 3x + \frac{1}{5}\sin 5x$ looks like

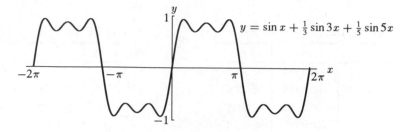

$$y = \sin x + \tfrac{1}{3}\sin 3x + \tfrac{1}{5}\sin 5x$$

(b) Following the pattern, we add the term $\tfrac{1}{7}\sin 7x$.

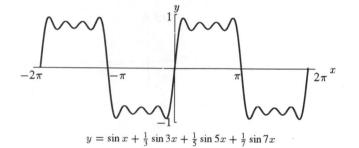

$$y = \sin x + \tfrac{1}{3}\sin 3x + \tfrac{1}{5}\sin 5x + \tfrac{1}{7}\sin 7x$$

(c) The equation is

$$f(x) = \begin{cases} \vdots & \vdots \\ 1 & -2\pi \le x < -\pi \\ -1 & -\pi \le x < 0 \\ 1 & 0 \le x < \pi \\ -1 & \pi \le x < 2\pi \\ \vdots & \vdots \end{cases}$$

The square wave function is not continuous at $x = 0,\ \pm\pi,\ \pm 2\pi, \ldots$

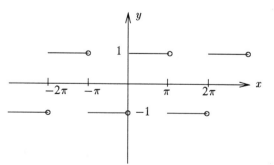

7. First,

$$a_0 = \frac{1}{2\pi} \int_{-\pi}^{\pi} f(x)\,dx = \frac{1}{2\pi}\left[\int_{-\pi}^{0} -x\,dx + \int_{0}^{\pi} x\,dx\right] = \frac{1}{2\pi}\left[-\frac{x^2}{2}\Big|_{-\pi}^{0} + \frac{x^2}{2}\Big|_{0}^{\pi}\right] = \frac{\pi}{2}.$$

To find the a_i's, we use the integral table. For $n \geq 1$,

$$
\begin{aligned}
a_n &= \frac{1}{\pi}\int_{-\pi}^{\pi} f(x)\cos(nx)\,dx = \frac{1}{\pi}\left[\int_{-\pi}^{0} -x\cos(nx)\,dx + \int_{0}^{\pi} x\cos(nx)\,dx\right]\\
&= \frac{1}{\pi}\left[\left(-\frac{x}{n}\sin(nx) - \frac{1}{n^2}\cos(nx)\right)\Big|_{-\pi}^{0}\right.\\
&\qquad \left. + \left(\frac{x}{n}\sin(nx) + \frac{1}{n^2}\cos(nx)\right)\Big|_{0}^{\pi}\right]\\
&= \frac{1}{\pi}\left(-\frac{1}{n^2} + \frac{1}{n^2}\cos(-n\pi) + \frac{1}{n^2}\cos(n\pi) - \frac{1}{n^2}\right)\\
&= \frac{2}{\pi n^2}(\cos n\pi - 1)
\end{aligned}
$$

Thus, $a_1 = -\frac{4}{\pi}$, $a_2 = 0$, and $a_3 = -\frac{4}{9\pi}$.

To find the b_i's, note that $f(x)$ is even, so for $n \geq 1$, $f(x)\sin(nx)$ is odd. Thus, $\int_{-\pi}^{\pi} f(x)\sin(nx) = 0$, so all the b_i's are 0.

Thus $F_1 = F_2 = \frac{\pi}{2} - \frac{4}{\pi}\cos x$, $F_3 = \frac{\pi}{2} - \frac{4}{\pi}\cos x - \frac{4}{9\pi}\cos 3x$.

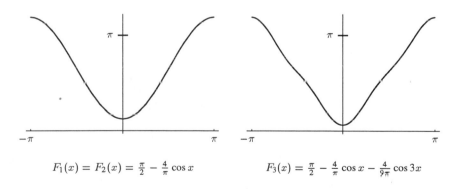

$F_1(x) = F_2(x) = \frac{\pi}{2} - \frac{4}{\pi}\cos x$ $\qquad\qquad$ $F_3(x) = \frac{\pi}{2} - \frac{4}{\pi}\cos x - \frac{4}{9\pi}\cos 3x$

9. First, we find a_0.

$$a_0 = \frac{1}{2\pi}\int_{-\pi}^{\pi} x^2\,dx = \frac{1}{2\pi}\left(\frac{x^3}{3}\Big|_{-\pi}^{\pi}\right) = \frac{\pi^2}{3}.$$

To find a_n, $n \geq 1$, we use the integral table (formulas III- 15 and III- 16)

$$a_n = \frac{1}{\pi} \int_{-\pi}^{\pi} x^2 \cos nx \, dx = \frac{1}{\pi} \left[\frac{x^2}{n} \sin(nx) + \frac{2x}{n^2} \cos(nx) - \frac{2}{n^3} \sin(nx) \right] \Bigg|_{-\pi}^{\pi}$$

$$= \frac{1}{\pi} \left[\frac{2\pi}{n^2} \cos(n\pi) + \frac{2\pi}{n^2} \cos(-n\pi) \right]$$

$$= \frac{4}{n^2} \cos(n\pi).$$

Again, $\cos(n\pi) = (-1)^n$ for all integers n, so $a_n = (-1)^n \frac{4}{n^2}$. Note that

$$b_n = \frac{1}{\pi} \int_{-\pi}^{\pi} x^2 \sin nx \, dx.$$

x^2 is an even function, and $\sin nx$ is odd, so $x^2 \sin nx$ is odd. Thus $\int_{-\pi}^{\pi} x^2 \sin nx \, dx = 0$, and $b_n = 0$ for all n.

We deduce that the nth Fourier polynomial for f (where $n \geq 1$) is

$$F_n(x) = \frac{\pi^2}{3} + \sum_{i=1}^{n} (-1)^i \frac{4}{i^2} \cos(ix).$$

In particular, we have the following graphs:

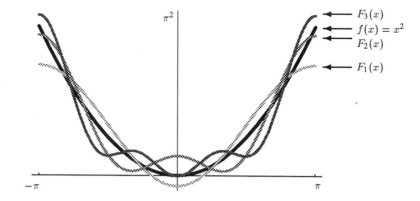

13. We have $f(x) = x, 0 \le x < 1$. Let $t = 2\pi x - \pi$. Notice that as x varies from 0 to 1, t varies from $-\pi$ to π. Thus if we rewrite the function in terms of t, we can find the Fourier series in terms of t in the usual way. To do this, let $g(t) = f(x) = x = \frac{t+\pi}{2\pi}$ on $-\pi \le t < \pi$. We now find the fourth degree Fourier polynomial for g.

$$a_o = \frac{1}{2\pi} \int_{-\pi}^{\pi} g(t)\, dt = \frac{1}{2\pi} \int_{-\pi}^{\pi} \frac{t+\pi}{2\pi}\, dt = \frac{1}{(2\pi)^2} \left(\frac{t^2}{2} + \pi t \right) \Big|_{-\pi}^{\pi} = \frac{1}{2}$$

Notice, a_0 is the average value of both f and g. For $n \ge 1$

$$a_n = \frac{1}{\pi} \int_{-\pi}^{\pi} \frac{t+\pi}{2\pi} \cos(nt)\, dt = \frac{1}{2\pi^2} \int_{-\pi}^{\pi} (t\cos(nt) + \pi\cos(nt))\, dt$$

$$= \frac{1}{2\pi^2} \left[\frac{t}{n} \sin(nt) + \frac{1}{n^2}\cos(nt) + \frac{\pi}{n}\sin(nt) \right] \Big|_{-\pi}^{\pi}$$

$$= 0.$$

$$b_n = \frac{1}{\pi} \int_{-\pi}^{\pi} \frac{t+\pi}{2\pi} \sin(nt)\, dt = \frac{1}{2\pi^2} \int_{-\pi}^{\pi} (t\sin(nt) + \pi\sin(nt))\, dt$$

$$= \frac{1}{2\pi^2} \left[-\frac{t}{n}\cos(nt) + \frac{1}{n^2}\sin(nt) - \frac{\pi}{n}\cos(nt) \right] \Big|_{-\pi}^{\pi}$$

$$= \frac{1}{2\pi^2} \left(-\frac{4\pi}{n}\cos(\pi n) \right) = -\frac{2}{\pi n}\cos(\pi n) = \frac{2}{\pi n}(-1)^{n+1}.$$

We get the integrals for a_n and b_n using the integral table (formulas III- 15 and III- 16).

Thus, the Fourier polynomial of degree 4 for g is:

$$G_4(t) = \frac{1}{2} + \frac{2}{\pi}\sin t - \frac{1}{\pi}\sin 2t + \frac{2}{3\pi}\sin 3t - \frac{1}{2\pi}\sin 4t$$

Now, since $g(t) = f(x)$, the Fourier polynomial of degree 4 for f can be found by replacing t in terms of x again. Thus,

$$F_4(x) = \frac{1}{2} + \frac{2}{\pi}\sin(2\pi x - \pi) - \frac{1}{\pi}\sin(4\pi x - 2\pi) + \frac{2}{3\pi}\sin(6\pi x - 3\pi) - \frac{1}{2\pi}\sin(8\pi x - 4\pi).$$

Now, using the fact that $\sin(x - \pi) = -\sin x$ and $\sin(x - 2\pi) = \sin x$, etc., we have:

$$F_4(x) = \frac{1}{2} - \frac{2}{\pi}\sin(2\pi x) - \frac{1}{\pi}\sin(4\pi x) - \frac{2}{3\pi}\sin(6\pi x) - \frac{1}{2\pi}\sin(8\pi x).$$

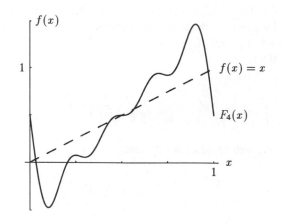

17.   The energy of the function $f(x)$ is

$$E = \frac{1}{\pi} \int_{-\pi}^{\pi} (f(x))^2 \, dx = \frac{1}{\pi} \int_{-\pi}^{\pi} x^2 \, dx = \frac{1}{3\pi} x^3 \Big|_{-\pi}^{\pi}$$

$$= \frac{1}{3\pi}(\pi^3 - (-\pi^3)) = \frac{2\pi^3}{3\pi} = \frac{2}{3}\pi^2 = 6.57974$$

From Problem 7, we know all the b_i's are 0 and $a_0 = \frac{\pi}{2}$, $a_1 = -\frac{4}{\pi}$, $a_2 = 0$, $a_3 = -\frac{4}{9\pi}$. Therefore the energy in the constant term and first three harmonics is

$$A_0^2 + A_1^2 + A_2^2 + A_3^2 = 2a_0^2 + a_1^2 + a_2^2 + a_3^2$$

$$= 2\left(\frac{\pi^2}{4}\right) + \frac{16}{\pi^2} + 0 + \frac{16}{81\pi^2} = 6.57596$$

which means that they contain $\dfrac{6.57596}{6.57974} = 0.99942 \approx 99.942\%$ of the total energy

21.   (a)

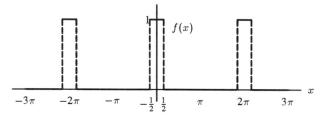

The energy of the pulse train f is

$$E = \frac{1}{\pi} \int_{-\pi}^{\pi} (f(x))^2 \, dx = \frac{1}{\pi} \int_{-1/2}^{1/2} 1^2 \, dx = \frac{1}{\pi}\left(\frac{1}{2} - \left(-\frac{1}{2}\right)\right) = \frac{1}{\pi}.$$

Next, find the Fourier coefficients:

$$a_0 = \text{average value of } f \text{ on } [-\pi, \pi] = \frac{1}{2\pi}(\text{ Area}) = \frac{1}{2\pi}(1) = \frac{1}{2\pi}$$

$$a_k = \frac{1}{\pi} \int_{-\pi}^{\pi} f(x) \cos kx \, dx = \frac{1}{\pi} \int_{-1/2}^{1/2} \cos kx \, dx = \frac{1}{k\pi} \sin kx \Big|_{-1/2}^{1/2}$$

$$= \frac{1}{k\pi}\left(\sin\left(\frac{k}{2}\right) - \sin\left(-\frac{k}{2}\right)\right) = \frac{1}{k\pi}\left(2\sin\left(\frac{k}{2}\right)\right)$$

$$b_k = \frac{1}{\pi} \int_{-\pi}^{\pi} f(x) \sin kx \, dx = \frac{1}{\pi} \int_{-1/2}^{1/2} \sin kx \, dx = -\frac{1}{k\pi} \cos kx \Big|_{-1/2}^{1/2}$$

$$= -\frac{1}{k\pi}\left(\cos\left(\frac{k}{2}\right) - \cos\left(-\frac{k}{2}\right)\right) = \frac{1}{k\pi}(0) = 0$$

The energy of f contained in the constant term is

$$A_0^2 = 2a_0^2 = 2\left(\frac{1}{2\pi}\right)^2 = \frac{1}{2\pi^2}$$

which is

$$\frac{A_0^2}{E} = \frac{1/2\pi^2}{1/\pi} = \frac{1}{2\pi} \approx 0.159155 = 15.9155\% \quad \text{of the total.}$$

The fraction of energy contained in the first harmonic is

$$\frac{A_1^2}{E} = \frac{a_1^2}{E} = \frac{\left(\frac{2\sin\frac{1}{2}}{\pi}\right)^2}{\frac{1}{\pi}} \approx 0.292653.$$

The fraction of energy contained in both the constant term and the first harmonic together is

$$\frac{A_0^2}{E} + \frac{A_1^2}{E} \approx 0.159155 + 0.292653 = 0.451808\%.$$

(b) The formula for the energy of the kth harmonic is

$$A_k^2 = a_k^2 + b_k^2 = \left(\frac{2\sin\frac{k}{2}}{k\pi}\right)^2 + 0^2 = \frac{4\sin^2\frac{k}{2}}{k^2\pi^2}$$

By graphing it as a continuous function for $k \geq 1$, we see its overall behavior as k gets larger. The energy spectrum for the first five terms is graphed below, as well.

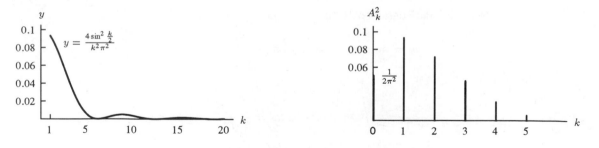

(c) The constant term and the first five harmonics are needed to capture 90% of the energy of f. This was determined by adding the fractions of energy of f contained in each harmonic until the sum reached at least 90% of the total energy of f:

$$\frac{A_0^2}{E} + \frac{A_1^2}{E} + \frac{A_2^2}{E} + \frac{A_3^2}{E} + \frac{A_4^2}{E} + \frac{A_5^2}{E} \approx 90.1995\%$$

(d) $F_5(x) = \frac{1}{2\pi} + \frac{2\sin(\frac{1}{2})}{\pi}\cos x + \frac{\sin 1}{\pi}\cos 2x + \frac{2\sin(\frac{3}{2})}{3\pi}\cos 3x + \frac{\sin 2}{2\pi}\cos 4x + \frac{2\sin(\frac{5}{2})}{5\pi}\cos 5x$

25. By formula II-12 of the integral table,

$$\int_{-\pi}^{\pi} \sin kx \cos mx \, dx$$

$$= \frac{1}{m^2 - k^2} \left(m\sin(kx)\sin(mx) + k\cos(kx)\cos(mx) \right)\Big|_{-\pi}^{\pi}$$

$$= \frac{1}{m^2 - k^2} \Big[m\sin(k\pi)\sin(m\pi) + k\cos(k\pi)\cos(m\pi)$$

$$- m\sin(-k\pi)\sin(-m\pi) - k\cos(-k\pi)\cos(-m\pi) \Big]$$

Since k and m are positive integers, $\sin(k\pi) = \sin(m\pi) = \sin(-k\pi) = \sin(-m\pi) = 0$. Also, $\cos(k\pi) = \cos(-k\pi)$ since $\cos x$ is even. Thus this expression reduces to 0. [Note: since $\sin kx \cos mx$ is odd, so $\int_{-\pi}^{\pi} \sin kx \cos mx \, dx$ must be 0.]

27. We make the substitution $u = mx$, $dx = \frac{1}{m}du$. Then

$$\int_{-\pi}^{\pi} \cos^2 mx \, dx = \frac{1}{m} \int_{u=-m\pi}^{u=m\pi} \cos^2 u \, du$$

By Formula IV- 18 of the integral table, this equals

$$\frac{1}{m}\left[\frac{1}{2}\cos u \sin u\right]\Bigg|_{-m\pi}^{m\pi} + \frac{1}{m}\frac{1}{2}\int_{-m\pi}^{m\pi} 1 \, du$$

$$= 0 + \frac{1}{2m}u\Bigg|_{-m\pi}^{m\pi} = \frac{1}{2m}u\Bigg|_{-m\pi}^{m\pi}$$

$$= \frac{1}{2m}(2m\pi) = \pi.$$

29. The easiest way to do this is to use Problem 27.

$$\int_{-\pi}^{\pi} \sin^2 mx \, dx = \int_{-\pi}^{\pi}(1 - \cos^2 mx) \, dx = \int_{-\pi}^{\pi} dx - \int_{-\pi}^{\pi} \cos^2 mx \, dx$$

$$= 2\pi - \pi \quad \text{using Problem 27}$$

$$= \pi.$$

SOLUTIONS TO REVIEW PROBLEMS FOR CHAPTER TEN

1. Substituting $y = t^2$ in $\sin y = y - \dfrac{y^3}{3!} + \dfrac{y^5}{5!} - \dfrac{y^7}{7!} + \cdots$ gives

$$\sin t^2 = t^2 - \frac{t^6}{3!} + \frac{t^{10}}{5!} - \frac{t^{14}}{7!} + \cdots$$

5.

$$\frac{a}{a+b} = \frac{a}{a(1+\frac{b}{a})} = \left(1+\frac{b}{a}\right)^{-1} = 1 - \frac{b}{a} + \left(\frac{b}{a}\right)^2 - \left(\frac{b}{a}\right)^3 + \cdots$$

9. $\ln x \approx \ln 2 + \dfrac{1}{2}(x-2) - \dfrac{1}{8}(x-2)^2$

13. (a) $f(t) = te^t$.

Use the Taylor expansion for e^t :

$$f(t) = t \left(1 + t + \frac{t^2}{2!} + \frac{t^3}{3!} + \cdots \right)$$

$$= t + t^2 + \frac{t^3}{2!} + \frac{t^4}{3!} + \cdots$$

(b)

$$\int_0^x f(t)\, dt = \int_0^x te^t\, dt = \int_0^x \left(t + t^2 + \frac{t^3}{2!} + \frac{t^4}{3!} + \cdots \right) dt$$

$$= \frac{t^2}{2} + \frac{t^3}{3} + \frac{t^4}{4 \cdot 2!} + \frac{t^5}{5 \cdot 3!} + \cdots \Big|_0^x$$

$$= \frac{x^2}{2} + \frac{x^3}{3} + \frac{x^4}{4 \cdot 2!} + \frac{x^5}{5 \cdot 3!} + \cdots$$

(c) Substitute $x = 1$:

$$\int_0^1 te^t\, dt = \frac{1}{2} + \frac{1}{3} + \frac{1}{4 \cdot 2!} + \frac{1}{5 \cdot 3!} + \cdots$$

In the integral above, to integrate by parts, let $u = t$, $dv = e^t\, dt$, so $du = dt$, $v = e^t$.

$$\int_0^1 te^t\, dt = te^t \Big|_0^1 - \int_0^1 e^t\, dt = e - (e - 1) = 1$$

Hence

$$\frac{1}{2} + \frac{1}{3} + \frac{1}{4 \cdot 2!} + \frac{1}{5 \cdot 3!} + \cdots = 1.$$

17. (a) To find when V takes on its minimum values, set $\frac{dV}{dr} = 0$. So

$$-V_0 \frac{d}{dr} \left(2 \left(\frac{r_0}{r} \right)^6 - \left(\frac{r_0}{r} \right)^{12} \right) = 0$$

$$-V_0 \left(-12 r_0^6 r^{-7} + 12 r_0^{12} r^{-13} \right) = 0$$

$$12 r_0^6 r^{-7} = 12 r_0^{12} r^{-13}$$

$$r_0^6 = r^6$$

$$r = r_0$$

Rewriting $V'(r)$ as $\dfrac{12 r_0^6 V_0}{r^7} \left(1 - \left(\dfrac{r_0}{r} \right)^6 \right)$, we see that $V'(r) > 0$ for $r > r_0$ and $V'(r) < 0$ for $r < r_0$. Thus, $V = -V_0(2(1)^6 - (1)^{12}) = -V_0$ is a minimum.

(Note: We discard the negative root $-r_0$ since the distance r must be positive.)

(b)

$$V(r) = -V_0 \left(2 \left(\frac{r_0}{r} \right)^6 - \left(\frac{r_0}{r} \right)^{12} \right) \qquad\qquad V(r_0) = -V_0$$
$$V'(r) = -V_0(-12r_0^6 r^{-7} + 12r_0^{12} r^{-13}) \qquad V'(r_0) = 0$$
$$V''(r) = -V_0(84r_0^6 r^{-8} - 156r_0^{12} r^{-14}) \qquad V''(r_0) = 72V_0 r_0^{-2}$$

The Taylor series is thus:

$$V(r) = -V_0 + 72V_0 r_0^{-2} \cdot (r - r_0)^2 \cdot \frac{1}{2} + \cdots$$

(c) The difference between V and its minimum value $-V_0$ is

$$V - (-V_0) = 36V_0 \frac{(r - r_0)^2}{r_0^2} + \cdots$$

which is approximately proportional to $(r - r_0)^2$ since terms containing higher powers of $(r - r_0)$ have relatively small values for r near r_0.

(d) From part (a) we know that $\frac{dV}{dr} = 0$ when $r = r_0$, hence $F = 0$ when $r = r_0$. Since, if we discard powers of $(r - r_0)$ higher than the second,

$$V(r) \approx -V_0 \left(1 - 36\frac{(r - r_0)^2}{r_0^2} \right)$$

giving

$$F = -\frac{dV}{dr} \approx 72 \cdot \frac{r - r_0}{r_0^2}(-V_0) = -72V_0 \frac{r - r_0}{r_o^2}.$$

So F is approximately proportional to $(r - r_0)$.

21. (a) Since $g^{(k)}(0)$ exists for all $k \geq 0$, and $g'(0) = 0$ because g has a critical point at $x = 0$. For $n \geq 2$,

$$g(x) \approx P_n(x) = g(0) + \frac{g''(0)}{2!}x^2 + \frac{g'''(0)}{3!}x^3 + \cdots + \frac{g^{(n)}(0)}{n!}x^n.$$

(b) The Second Derivative test says that if $g''(0) > 0$, then 0 is a local minimum and if $g''(0) < 0$, 0 is a local maximum.

(c) Let $n = 2$. Then $P_2(x) = g(0) + \frac{g''(0)}{2!}x^2$. So, for x near 0,

$$g(x) - g(0) \approx \frac{g''(0)}{2!}x^2.$$

If $g''(0) > 0$, then $g(x) - g(0) \geq 0$, as long as x stays near 0. In other words, there exists a small interval around $x = 0$ such that for any x in this interval $g(x) \geq g(0)$. So $g(0)$ is a local minimum.

The case when $g''(0) < 0$ is treated similarly; then $g(0)$ is a local maximum.

25. We assume the interest rate is i per year. Then

$$\text{Present value of first coupon} = \frac{50}{1+i}$$

$$\text{Present value of second coupon} = \frac{50}{(1+i)^2}$$

Thus

$$\text{Total present value} = \underbrace{\frac{50}{1+i} + \frac{50}{(1+i)^2} + \cdots + \frac{50}{(1+i)^{10}}}_{\text{coupons}} + \underbrace{\frac{1000}{(1+i)^{10}}}_{\text{principal}}$$

$$= \frac{50}{1+i}\left(1 + \frac{1}{1+i} + \cdots + \frac{1}{(1+i)^9}\right) + \frac{1000}{(1+i)^{10}}$$

$$= \frac{50}{1+i}\left(\frac{1 - \left(\frac{1}{1+i}\right)^{10}}{1 - \frac{1}{1+i}}\right) + \frac{1000}{(1+i)^{10}}$$

$$= \frac{50}{1+i}\left(\frac{1 - \left(\frac{1}{1+i}\right)^{10}}{1 - \frac{1}{1+i}}\right)\left(\frac{(1+i)^{10}}{(1+i)^{10}}\right) + \frac{1000}{(1+i)^{10}}$$

$$= \frac{50}{1+i}\left(\frac{((1+i)^{10} - 1)}{(1+i)^{10}}\right)\frac{1+i}{(1+i-1)} + \frac{1000}{(1+i)^{10}}$$

$$= \frac{50((1+i)^{10} - 1)}{i(1+i)^{10}} + \frac{1000}{(1+i)^{10}}$$

Since we know the present value is \$950, we want to solve for i giving

$$\frac{50}{i}\frac{((1+i)^{10} - 1)}{(1+i)^{10}} + \frac{1000}{(1+i)^{10}} = 950$$

This equation cannot be solved analytically for i. However graphing the function

$$f(i) = \frac{50((1+i)^{10} - 1)}{i(1+i)^{10}} + \frac{1000}{(1+i)^{10}} - 950$$

and zooming on the zero shows that $f(i) = 0$ when $i \approx 0.057$, so an interest rate of about 5.7% per year gives a present value of \$950.

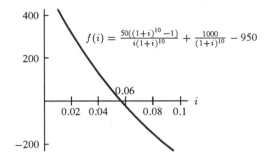

$$f(i) = \frac{50((1+i)^{10}-1)}{i(1+i)^{10}} + \frac{1000}{(1+i)^{10}} - 950$$

29. Let $t = 2\pi x - \pi$. Then, $g(t) = f(x) = e^{2\pi x} = e^{t+\pi}$. Notice that as x varies from 0 to 1, t varies from $-\pi$ to π. Thus, we can find the Fourier coefficients for $g(t)$:

$$a_o = \frac{1}{2\pi}\int_{-\pi}^{\pi} g(t)dt = \frac{1}{2\pi}\int_{-\pi}^{\pi} e^{t+\pi}dt = \frac{1}{2\pi}e^{t+\pi}\Big|_{-\pi}^{\pi} = \frac{e^{2\pi}-1}{2\pi}$$

$$a_n = \frac{1}{\pi}\int_{-\pi}^{\pi} e^{t+\pi}\cos(nt)dt = \frac{e^{\pi}}{\pi}\int_{-\pi}^{\pi} e^t\cos(nt)dt$$

Using the integral table, Formula II- 8, yields:

$$= \frac{e^{\pi}}{\pi}\frac{1}{n^2+1}e^t(\cos(nt)+n\sin(nt))\Big|_{-\pi}^{\pi}$$

$$= \frac{e^{\pi}}{\pi}\frac{1}{n^2+1}(e^{\pi}-e^{-\pi})(\cos(n\pi))$$

$$= \frac{(e^{2\pi}-1)}{\pi}\frac{(-1)^n}{n^2+1}$$

$$b_n = \frac{1}{\pi}\int_{-\pi}^{\pi} e^{t+\pi}\sin(nt)dt = \frac{e^{\pi}}{\pi}\int_{-\pi}^{\pi} e^t\sin(nt)dt$$

Again, using the integral table, Formula II- 9, yields:

$$= \frac{e^{\pi}}{\pi}\frac{1}{n^2+1}e^t(\sin(nt)-n\cos(nt))\Big|_{-\pi}^{\pi}$$

$$= -\frac{e^{\pi}}{\pi}\frac{n}{n^2+1}(e^{\pi}-e^{-\pi})\cos(n\pi)$$

$$= \frac{(e^{2\pi}-1)}{\pi}\frac{(-1)^{n+1}n}{n^2+1}$$

Thus, after factoring a bit, we get:

$$G_3(t) = \frac{e^{2\pi}-1}{\pi}\left(\frac{1}{2} - \frac{1}{2}\cos t + \frac{1}{2}\sin t + \frac{1}{5}\cos 2t - \frac{2}{5}\sin 2t - \frac{1}{10}\cos 3t + \frac{3}{10}\sin 3t\right)$$

Now, we substitute x back in for t:

$$F_3(x) = \frac{e^{2\pi} - 1}{\pi}(\frac{1}{2} - \frac{1}{2}\cos(2\pi x - \pi) + \frac{1}{2}\sin(2\pi x - \pi) + \frac{1}{5}\cos(4\pi x - 2\pi)$$

$$-\frac{2}{5}\sin(4\pi x - 2\pi) - \frac{1}{10}\cos(6\pi x - 3\pi) + \frac{3}{10}\sin(6\pi x - 3\pi))$$

Recalling that $\cos(x - \pi) = -\cos x$, $\sin(x - \pi) = -\sin x$, $\cos(x - 2\pi) = \cos x$, and $\sin(x - 2\pi) = \sin x$, we have:

$$F_3(x) = \frac{e^{2\pi} - 1}{\pi}\left(\frac{1}{2} + \frac{1}{2}\cos 2\pi x - \frac{1}{2}\sin 2\pi x + \frac{1}{5}\cos 4\pi x - \frac{2}{5}\sin 4\pi x\right.$$

$$\left. + \frac{1}{10}\cos 6\pi x - \frac{3}{10}\sin 6\pi x\right)$$

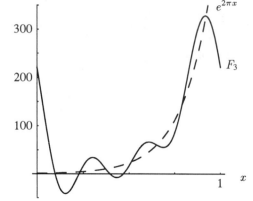

33. Since $g(x) = f(x + c)$, we have that $[g(x)]^2 = [f(x + c)]^2$, so g^2 is f^2 shifted horizontally by c. Since f has period 2π, so does f^2 and g^2. If you think of the definite integral as an area, then because of the periodicity, integrals of f^2 over any interval of length 2π have the same value. So

$$\int_{-\pi}^{\pi} (f(x))^2\, dx = \int_{-\pi+c}^{\pi+c} (f(x))^2\, dx$$

Now we know that

$$\text{Energy of } g = \frac{1}{\pi}\int_{-\pi}^{\pi} (g(x))^2\, dx$$

$$= \frac{1}{\pi}\int_{-\pi}^{\pi} (f(x + c))^2\, dx.$$

Using the substitution $t = x + c$, we transform the integral on the right, obtaining

$$= \frac{1}{\pi} \int_{-\pi+c}^{\pi+c} (f(t))^2 \, dt$$

$$= \frac{1}{\pi} \int_{-\pi}^{\pi} (f(t))^2 \, dt \quad \text{from above equation}$$

$$= \text{Energy of } f$$

APPENDIX

A SOLUTIONS

1. The graph is

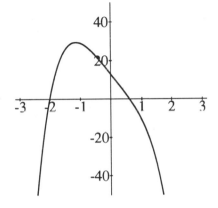

   (a)   The range appears to be $y \leq 30$.

   (b)   The function has two zeros.

5.   The largest root is at about 2.5.

9.   The root occurs between 1.1 and 1.2, at about 1.15.

13.   (a)   Only one real zero, at about $x = -1.15$.

      (b)   Three real zeros: at $x = 1$, and at about $x = 1.41$ and $x = -1.41$.

17.   (a)   Since f is continuous, there must be one zero between $\theta = 1.4$ and $\theta = 1.6$, and another between $\theta = 1.6$ and $\theta = 1.8$. These are the only clear cases. We might also want to investigate the interval $0.6 \leq \theta \leq 0.8$ since $f(\theta)$ takes on values close to zero on at least part of this interval. Now, $\theta = 0.7$ is in this interval, and $f(0.7) = -0.01 < 0$, so f changes sign twice between $\theta = 0.6$ and $\theta = 0.8$ and hence has two zeros on this interval (assuming f is not *really* wiggly here, which it's not). There are a total of 4 zeros.

      (b)   As an example, we find the zero of f between $\theta = 0.6$ and $\theta = 0.7$. $f(0.65)$ is positive; $f(0.66)$ is negative. So this zero is contained in $[0.65, 0.66]$. The other zeros are contained in the intervals $[0.72, 0.73]$, $[1.43, 1.44]$, and $[1.7, 1.71]$.

      (c)   You've found all the zeros. A picture will confirm this; see Figure A.1.

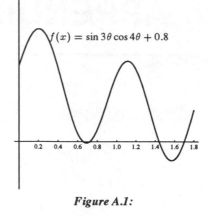

Figure A.1:

B   SOLUTIONS

1.   Yes, there is no break in the graph of $f(x)$ although it does have a 'corner' at $x = 0$.

5.

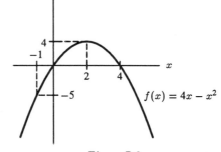

Figure B.2

Bounded and $-5 \leq f(x) \leq 4$.

C   SOLUTIONS

1. $(1,0)$

5. $\left(\frac{5\sqrt{3}}{2}, -\frac{5}{2}\right)$

9. $r = \sqrt{0^2 + 2^2} = 2, \quad \theta = \pi/2$.

13. $r = \sqrt{(0.2)^2 + (-0.2)^2} = 0.28$.
$\tan\theta = 0.2/(-0.2) = -1$. Since the point is in the fourth quadrant, $\theta = 7\pi/4$. (Alternatively $\theta = -\pi/4$.)

D   SOLUTIONS

1. $2e^{\frac{i\pi}{2}}$

5. $0e^{i\theta}$, for any θ.

9. $-3 - 4i$

13. $\frac{1}{4} - \frac{9i}{8}$

17. $5^3(\cos\frac{3\pi}{2} + i\sin\frac{3\pi}{2}) = -125i$

21. One value of $\sqrt[3]{i}$ is $\sqrt[3]{e^{i\frac{\pi}{2}}} = (e^{i\frac{\pi}{2}})^{\frac{1}{3}} = e^{i\frac{\pi}{6}} = \cos\frac{\pi}{6} + i\sin\frac{\pi}{6} = \frac{\sqrt{3}}{2} + \frac{i}{2}$

25. One value of $(-4+4i)^{2/3}$ is $[\sqrt{32}e^{i\frac{3\pi}{4}}]^{2/3} = (\sqrt{32})^{2/3}e^{i\frac{\pi}{2}} = 2^{\frac{10}{3}}\cos\frac{\pi}{2} + i2^{\frac{10}{3}}\sin\frac{\pi}{2} = 8i\sqrt[3]{2}$

29. Substituting $A_1 = 2 - A_2$ into the second equation gives

$$(1-i)(2-A_2) + (1+i)A_2 = 0$$

so

$$2iA_2 = -2(1-i)$$
$$A_2 = \frac{-(1-i)}{i} = \frac{-i(1-i)}{i^2} = i(1-i) = 1+i$$

Therefore $A_1 = 2 - (1+i) = 1 \quad i$.

33. True, since $\sqrt{a}$ is real for all $a \geq 0$.

37. True. We can write any nonzero complex number z as $re^{i\beta}$, where r and β are real numbers with $r > 0$. Since $r > 0$, we can write $r = e^c$ for some real number c. Therefore, $z = re^{i\beta} = e^c e^{i\beta} = e^{c+i\beta} = e^w$ where $w = c + i\beta$ is a complex number.

41. Using Euler's formula, we have:

$$e^{i(2\theta)} = \cos 2\theta + i\sin 2\theta$$

On the other hand,

$$e^{i(2\theta)} = (e^{i\theta})^2 = (\cos\theta + i\sin\theta)^2 = (\cos^2\theta - \sin^2\theta) + i(2\cos\theta\sin\theta)$$

Equating real parts, we find

$$\cos 2\theta = \cos^2\theta - \sin^2\theta.$$